VGM Opportunities Series

OPPORTUNITIES IN
TOOL AND DIE
CAREERS

George A. Dudzinski, Sr.

Foreword by
Matthew B. Coffey
President
National Tooling and Machining Association

VGM Career Horizons
a division of *NTC Publishing Group*
Lincolnwood, Illinois USA

Cover Photo Credits:

Front cover: upper left, Greenville Technical College; upper right and lower right, Eric Brady photo, courtesy of New River Community College; lower left, Mike Bland photo, courtesy of Kankakee Community College.

Back cover: upper left and lower left, Eric Brady photo, courtesy of New River Community College; upper right, Mike Bland photo, courtesy of Kankakee Community College.

Library of Congress Cataloging-in-Publication Data
Dudzinski, George A.
    Opportunities in tool & die careers / George A. Dudzinski.
        p. cm. — (VGM opportunities series)
    ISBN 0-8442-4047-8 (hard) — ISBN 0-8442-4048-6 (soft)
    1. Tool and die industry—Vocational guidance—United States.
  I. Title. II. Title: Opportunities in tool and die careers.
  III. Series.
HD9703.U5D83 1993
621.9'0023'73—dc20   92-41901
                                                              CIP

Published by VGM Career Horizons, a division of NTC Publishing Group.
© 1993 by NTC Publishing Group, 4255 West Touhy Avenue,
Lincolnwood (Chicago), Illinois 60646-1975 U.S.A.

3 4 5 6 7 8 9 0 VP 9 8 7 6 5 4 3 2 1

ABOUT THE AUTHOR

George A. Dudzinski, Sr., was born and educated in the state of Connecticut and comes from a family of toolmakers. Following his father and brothers, who were all toolmakers, he attended the state vocational technical school and later became a toolmaker apprentice. In 1971, after nine years of working in the field, he enrolled at the Central Connecticut State College, taking courses in teaching, student relationships, and development. He then taught in the state trade schools for the CETA program. He taught unemployed workers skills that would qualify them for jobs in the field and developed training programs and shop layout for these students.

After twelve years of teaching he returned to the industry and started an in-house toolmaking program for one of the largest corporations today where he taught tool and die apprentices, theory, and on-the-job training. He is now employed in the United States Army and involved in logistics. He is active with helping other people in the trade.

FOREWORD

The most highly skilled and valued employees in manufacturing are tool and die makers, mold makers, precision machinists and precision sheet-metal model makers. They are the "surgeons of steel," and they make mass manufacturing possible.

You become one by an apprenticeship or on the job internship. This means you must seek employment with a job shop to get into the training, which is normally paid for by the employer. The ideal intern has graduated from high school with better than average grades, good-to-excellent math skills, and good communication and listening skills. A good candidate also has a strong mechanical aptitude and is punctual, precise, neat, and patient. Computer skills are becoming mandatory.

Tool and die making is a great career with constant demand. If you like making things using your brain and your hands, here is the opportunity to get a fully paid four- or five-year education while being paid, without a college education.

The work is interesting and varies every day, and you will be part of new product development throughout your career.

Matthew B. Coffey
President
National Tooling and Machining Association

Dedications for
Gary Lassy; Walter Dudzinski, Sr.; my wife, Wanda;
my children, Kim, Eli, David, Jacob, and George, Jr.
To my friends David, Gene, Rick, Lucian, Jimmy,
and Buster
To my new friends Carl, Nick, Larry, and Steve
Special thanks to Mr. Frank Danijovicks
(Thanks Frank)

CONTENTS

CHAPTER 1

THE AREAS COVERED BY TOOL AND DIE

MACHINERY USED IN THE TOOL AND DIE FIELD

Machines play an important part in the tool and die industry. They serve as a tool to guide workers in performing their job. Similar products that are precise and of high quality are the toolmaker's goal. The machines that you will encounter in the trade will be similar to those that we will discuss in this book. Although the field is rapidly modernizing with new technology and simple hands-on operations are being replaced with computer-operated machines, it is still important to understand the basic function of these machines.

We will discuss each machine in detail in the following pages of this chapter.

Drill Press Information

There are many varieties of drill presses, from the large floor models to the smaller bench models. Yet their functions serve the same purpose, which is to drill holes of any size and diameter.

There are many steps to drilling. Center drilling is the process by which a precise location for the hole is achieved. The next step is called drilling, which begins after you determine the size you require. You then drill that size in the metal. It is important to remember that a drill will put a hole in a piece of metal. However, you do not want to start the process with the actual size of the hole you want to drill. Instead you want to gradually bring it up to the size. Drills are known to "walk." This term is referred to often in the field. Once started the drill will only go in one direction which is usually down. However if you apply too much force the hole will turn crooked, or *walk*.

A newly drilled hole has a burr. This is raised metal which occurs on the top and bottom of the hole. You must make sure you remove this burr before any other procedures can be done. This process is called chamfering. The chamfering tool is a drill with a precise angle. It removes the raised metal which is called a burr or ruff edge. The chamfer has to be larger than your finished hole.

The next process involves the use of a tool called a reamer which is used to finalize the drilled hole. Reamed holes have an accurate finish and smooth surface. You must always drill your holes smaller than the finished hole so that the reamer will have material to remove.

Another procedure is called counterboring. This is when you want to enlarge an already existing hole to fit a bolt head or screw head inside the hole. The counterbore has a pilot in the front of it which is removable. To increase the size of your drilled hole it has a cutterlike milling tool that will increase the size of the existing hole to fit the screw head inside.

Another boring method allows you to simply use a boring head. This boring head is adjustable and normally moves in small increments of thousandths of an inch. It has a tool attached to it which can be moved outward to achieve a larger hole size.

Math is used in the drilling process. The formula for finding rpm (cutting speed) is the cutting speed multiplied by 12 and then divided by the diameter. This number is then multiplied by 3.1416. This is one of many formulas that you will use. All formulas can be found in the *Machinery's Handbook*, the toolmaker's bible.

The following safety list is very important when using the drill press:

- You should wear glasses at all times to avoid getting hit in the eye.
- Never wear loose clothes when working with a drill press.
- Do not stop the drill chucks by hand (see holding devices).

- Use vises and not your hand to hold pieces.
- Tie up loose hair.
- Run at a safe speed.
- Always keep your mind on your work.
- Never horse around when operating the drill press.
- Never leave the machine running and walk away from it.

Lathes

Lathes come in all sizes and can be hand driven or computer driven. A lathe is similar to a drill press placed on its side. A lathe also will make a majority of different metal shapes. A standard type of lathe can do many operations. One industry that makes use of the lathe is the automotive field. CAM shafts, crankshafts, valves, bearings, bearing casing headbolts, and transmission threaded shafts are a few of the products that can be made on a lathe. Lathe procedures are listed in the following pages.

Turning a piece of material is like peeling an apple with an electric knife because it involves one smooth cut. You can turn a part from left to right, right to left, or in and out. Everyone in the field calls it "tapering a piece of stock." Accurate sizing is what is sought after. This can be achieved by placing the material between centers or held in a collet or a chuck. Feeds and speeds (see milling machines) have to be figured out to achieve the best finish for your work. Tooling has to be figured out as well, which involves radius, angle, and, finish. A good tool with the correct angle will do this for you.

Tooling used for turning are made of tool steel (steel that is fire-hardened and used to cut softer metals) or tool steel with carbide inserts. Tool steel comes in different sizes which you must grind in order to do the operation. Carbide inserts are molded from the manufacture to given angles which are already ground. Insert tool holders come in different shapes and are either locked in a given slot or locked down with screws. Tool bits that are tool steel can be sharpened over and over again. However, carbide inserts can be resharpened but are usually replaced and the used one thrown out. The advantage of the insert is that it can cut faster and smoother and more material can be taken off at one time. The tool bit can be resharpened, but it takes much longer to set up again.

Threads are made on a lathe by using a threading tool. This tool is also tool steel and called a tool bit. Inserts are used in the operation for less time and faster running. The angles on a threading tool are 30° × 30° or 59° total is the common angle used in English threads. You must grind the tool bit in an angle using a center gage. This gage is shaped like an arrow. There are two angles on the side of it for large and small tools. You must also set the machine up to achieve the correct straightness. Inserts are made with the angle already ground in from the manufacturer. These inserts are thrown away just like the turning tool. The advantage in using the insert is that when it wears out, all you have to do is change it and minor adjustments are made. There are many varieties of holds in threading. Although most people think thread angles are a point, there are square threads as well.

Knurling is a procedure whereby a tool is pushed with a diamond design cutter forcing out the metal. The work is held in many ways but the best ways are in a chuck or between centers. There is another knurl which is called a straight line knurl. The holder used here normally has two matching rollers. When pushed in any direction, the tool forces out the material to a diamond or straight knurl cut. Knurls can be found on almost anything. Some items are screw heads, wristwatches, and fishing poles.

The following safety list is important when operating a lathe:

- Watch for flying objects from other machines.
- Remember to always wear glasses in the shop area.
- Remove the chuck key from the chuck when finished. (Students and toolmakers have a bad habit of leaving it in.)
- Never pull on long chips coming out of the machine.
- Look at what you are doing and do not daydream.
- Remember that when using cutting oil it may splash and may be extremely hot.
- Polishing parts in small and large machines is very hazardous if not done correctly.

- Stopping rotating chucks and parts in the machine should be done by the machine itself, not your hand.
- Always use the correct coolant for the material you are using. If the wrong coolant is used on some material, it will start a fire.
- Speeds and feeds are very important to know when drilling and turning.

Milling Machines

There are many types of milling machines used in the industries. Today the two basic ones that machinists and toolmakers use are the vertical and the horizontal milling machines. The milling machines can do almost anything you would like them to do, anything from peeling off steel at a light cut to the heaviest cut you can take at one time. The vertical milling machine can mill slots, make grooves, drill, tap a hole, fly cut, T-slot, angle cut, and much more. The milling machine moves on an axis—left to right, right to left, in and out, and up and down.

With the speeds and feeds machine, the manufacturer has a set chart that you can go by, but it is up to you to know what speed and feed to use. The charts that are available give you a rough idea of what the manufacturer recommends for your cutting speed (rpm) and feet per minute. The machine has many attachments that be be used. They are: collets, drill chucks, quick change spindles, boring heads, and slotting attachments. Some of the milling machines have a tracer control and some have the CAD CAM RS 232 down load, a computer-controlled blueprinting guidance system. They all work fine provided that you understand them and listen to your instructor or supervisor. All operations that are done on the vertical milling machine can be held within .0005 tenths of an inch. The machine has a table that is used to hold vises, and your work can be bolted to it. The cutter is in one position and the only thing that you move is the table. This utilizes the axis method.

Another milling machine is called the horizontal miller, and it is used for removing large amounts of material at one time. It is basically the same method as the vertical milling machine with some exceptions. Its head is horizontal and can cover more area then the vertical miller can. Some industries have computer systems to cut the time factor down to a minimum. This machine is used for the heavy-duty work that can not be done on the vertical miller.

Holding Devices

The horizontal millers have vises to lock and hold down parts of steel that you are milling. Basically there are four standard vises: the plain vise that locks on the table slots and is used for heavy cutting; the swivel vise that locks the same way but can be turned 360 degrees and is used for cutting angles, grooves, and slots (basically this vise is still one of the heavy cutting vises); the complex angle vise or toolmaker vise can be used for multiangle cutting (known as the two-plane compound angle); and the adjustable angle vise which can only be used for one plane (angle). The complex and the adjustable vises are used for light cutting only. There is one more type of vise the toolmakers use. It is called the cross slide table. It is a vise mounted to the table of the machine and can be positioned on one plane (angle). It has a small table on it and can be moved just like the milling machine table.

The rotary table is a holding device that is used when you need accurate cutting such as correct spacing, radius cutting, and dividing. The table can be turned 360 degrees and timed in minutes and even seconds. It is mounted on the milling machine table. You use this device to make circles, evenly spaced slots, grooves, and drilled holes. Pieces are mounted on it and held down the same way you lock in the vises.

An index head is used to cut gears, mill slots, and drill holes. The head has templates which are round and which have many holes in them. The holes are used to move the piece into a correct position as it is moved.

There are two types of chucks—the standard universal chuck and the rotary table type. The standard chuck has three jaws for holding a piece of any shape and is used for milling and drilling. The machine can be run at a fast or slow speed because of its strong holding power. The rotary chuck can do the same thing but it can turn 360 degrees at one-degree increments. It is also strong enough for heavy cuts. These chucks are also equipped with more jaws for holding power.

There are many tool holders used in this machine. One of them is called an arbor. It is long with a threaded end that has sleeves which slide on to it. The cutters that you would use have the same inside diameter as the diameter of the arbor. You can use one or more cutters on this shaft at the same time. The arbor is supported by the head of the machine and by an overhanging support that will hold the arbor steady. Like the vertical miller, you can use only one cutter at a time.

Sleeve and adapter holders are similar to a collet holder. They are smaller but serve the same purpose in milling. These holders can be set up for milling cutters, endmills, drills, drill chucks, and reamers to complete the job. Unlike the arbor, which is a supported shaft, these have the support in the head of the milling machine.

Measuring is a major factor here. You must have an excellent math background. You have to know what size cutter to use and where to locate it in the holders or the arbor.

There are different types of cutters that are used in the field. Shell mills cutters take the material off at one time. They can be used on the milling machine arbor or sleeve holders and can cut two surfaces at one time at a slow or fast speed.

Woodruff key cutters are used to put key slots on a piece of steel. You would see this in many items from weapon sights to crankshafts in engines. Some other cutters are

T-slot cutters. These are used in tables of milling machines or other machines that are used to hold in a bolt and hold your work on the table. Single and double angle cutters are used to put an angle on a piece of steel. They come one-sided or double-sided. They are used to produce V-slots. Convex and concave or milling cutters are used to cut inside and outside radius, respectively. The dovetail cutter is a cutter used to make two matching parts slide together. Common type endmills are similar to the ones you use on the vertical miller. Single cut, double cut, double end millers, tapered shank cutters, and slitting cutters are very thin and are used for cutting off material and cutting very narrow slots. Other important cutters are: stagger tooth, plain cutters, helical, heavy-duty plain cutters, and side-mill. Plain cutters are used to cut material off one surface. Helical cutters look like a screw and can run very fast because the cutter screws cut out the material like a thread. Carbide cutters come in inserts, throw away, and the solid type. The inserts can be used to take a lot of steel off at one time, and if you damage one you can replace it with setting up the machine again.

There are cutting speed formulas that you have to know. The following are the ones you use all the time: 1. rpm = 4 × the surface feet per minute divided by D. D is the diameter of the cutter in inches. Begin with the slowest and work up to the speed that will give you the best finish. A general rule is that it should be as coarse as possible and at the same time fine enough to have the cutter last a long time without breaking down. 2. A feed rate formula that you can use is $F = F \times T \times N$. This translates to: Feed rate (F) equals feed rate per tooth (F) times number of teeth (T) times rpm of cutter (N).

The following safety list is very important when using the holding devices and milling machines:

- Cutters should be handled with care because of the sharp edges.
- Pieces should be locked down in whatever holding device you are using.
- Millers run at a very fast rpm. If you have loose hair or hair that is very long, you have to be very careful that it doesn't get pulled into the spindle.
- Cutters that are used in this machine should be handled with care.
- Some milling machines have very heavy holding devices. You should take care in lifting them. Always have help with the large ones.
- Be very careful when using any carbide tooling. It will fly apart if hit the wrong way.
- Never forget the wrench that tightens down the shaft and tightens the tooling down. If not done properly damage could occur to the machine and possibly to yourself.

Electric Discharge Machine

An electric discharge machine (EDM) is a machine that uses electricity to make shapes such as impressions, indentions, holes, slots, grooves, and so forth. The process is achieved with current going through the head of the machine, through a piece of brass of any shape, and making contact with whatever is being manufactured or designed. The work is held down on a table similar to the millers with the same setup. Clamps and a ground wire are attached to the work. The head moves downward to a set location using dial indicators and actually burns into the metal, leaving a clean accurate size. This machine can be used to do a number of operations. It uses oil for coolant and to cool the work piece down while the probe is burning into the material. It is also very large and is about the same size as a vertical miller. The EDM is also able to make an impression in a die with the letters coming outward much like an engraving machine you may have seen in malls and jewelry stores. This machine engraves letters in a piece of material. Similarly, a machine like this is used to put patent numbers where raised letters are necessary in order to read the stock or patent number. In order to do this one must engrave the letters on the part or die and then put it in the EDM with the head locked in. The part that will have the impression will be locked on the table. The head then goes downward and the electrical charge burns into the material on the table and draws out the engraved part in the head. This makes the shape of whatever is engraved on that piece of material in the engraving machine. This process can be seen on telephones in the plastic housing of the main body.

There are various elements and parts to the EDM that are important to understand. The probes are made of brass. When using the EDM it is important to read your blueprint. Understanding your design is the first step in making your probe. The operation can be done on the lathe if you are making a round hole operation. If the blueprint calls for a square hex or oblong, the milling machines should be used. Coolant oil is pumped in for lubrication between the probe and the metal. With the probe and work under oil, the operation can then begin. With these precautions the metal will not catch on fire.

Measurements and the movements of the head on the EDM are similar to the vertical miller. A dial indicator attached to the left side of the head measures depth as the head moves downward. This is a sophisticated machine and is not intended for beginners. It is used widely in the aerospace sector of the industry. EDMs are now computerized, although many shops still use them manually.

The following safety list is important when operating the EDM:

- Oil coolant is one of the most important things in the machine. If the parts are not under the coolant when the process begins, an oil fire could start.
- Probes should have a good contact and not be loose.
- Alignment of the two mating parts should be closely set to assure safe contact.
- Always wear glasses when operating this machine because sometimes oil will shoot out when starting up the operation.

- Gages on the machine should be watched for down pressure loss.
- Polarity has to be going in the right direction and not in reverse.

Grinding Machines

There are different types of grinding machines and operations. The following will give you an idea of how many different machines you will encounter including the different wheel shapes and different operations.

Grinding is a procedure by which a smooth finish can be obtained. In toolmaking it is important to grind to a micro finish. The ability to procure a micro finish is acquired by experience. No one can just get on a grinding machine and grind. You must know the machine you want to use and what operation you need. Safety is a very important factor to observe when grinding metal. The wheels rotate at a high speed. If the wheel has not been tightened the right way, it will spin out of the spindle faster then you can turn the machine off. Some grinding machines give off a lot of heat. If a small amount of metal has to be removed there is no need for any substance to cool the metal. If a great amount of grinding is needed a machine with some type of coolant is necessary. Coolant is only used to keep the temperature down in the metal. If the temperature is not controlled you could expand your work piece and possibly crack the grinding wheel. There are many machines associated with grinding. The first one is the cylindrical grinder. This machine works with the small or large grinding wheel. It turns in one direction with the work piece turning in another. The work piece is mounted between centers with a holder locking it down on one end. The work moves from left to right and is fed inward into the grinding wheel. The machine has many speeds for the feeding of the work. The manufacturer will list the possible speeds on the machine itself. This machine is able to grind pieces from very thin to very large. On these types of machines there is coolant flowing on the part that is being ground to keep it cool. Parts of machines that could be ground on the machine are valves from an engine, parts for aerospace, crankshafts, jet engine parts, and many others.

The grinding wheel is mounted on a holder similar to the arbor. Grinding wheels come in different sizes.

Some of the operations that you can grind on this machine are: straight turning, taper turning, and form turning.

The next machine is the tool and cutter grinder. This machine grinds your cutting tools which are used on the milling machines, drill presses, and lathes. The head of this machine can move 360 degrees and can be adjusted for any type of cutter or tool that needs to be ground. This machine takes a lot of practice to run. The tool rest blades are for all different types of cutters. In this machine you actually have to make stops, which could be anything from taking a piece of sheet metal to making one out of one-inch square stock.

There are about sixteen or more operations to regrinding a cutter. These operations are different with every cutter or tool. When using the tool grinder, setups will vary; therefore, it is important to learn the proper procedure. There is safety involved with this machine that should be observed at all times. One important factor which promotes safety is to do a dry run. Toolmakers are the eyes and ears of a company. Make sure you know how a machine functions before you turn it on. Some toolmakers do not think in this area and many become seriously hurt. Many lose fingers or eyes and have scars for life. It is important to use your knowledge.

A surface grinder is a common machine used by all toolmakers today. It is a fast and quick machine that takes off steel in a moment, leaving a smooth finish to a piece of steel. A micro finish can easily be maintained. The machine has a table that moves left to right and right to left. You can easily grind steel and brass with the proper knowledge of the surface grinder.

The surface grinder can be automatic or manual according to various types available. The machine spindle turns at a high rate of speed and material is removed when the head moves downward into the piece of material at a rate of .0001 of an inch. The surface grinder is equipped with a magnetic chuck. It sits on top of the table and is locked down with T-bolts. The magnetic chuck is used to hold pieces of steel on it and keep them from moving. Pieces of steel smaller than the work piece are usually placed on all four sides. This will keep the work piece from moving, and it is sometimes referred to as "blocking." Stainless steel, bronze, and brass cannot be held on a magnetic chuck. These types of metals should be blocked with steel that can be held.

The centerless grinder is another machine that the toolmaker will use. This machine can grind a piece of material without holding it. The way the machine functions is that there is a large grinding wheel turning down on the the work and a regulating wheel that is turning upward. This second wheel usually is to the right side of the work piece. This second wheel differs structurally from the grinding wheel. Its primary function is to hold the work against the large grinding wheel that is grinding the work piece. Pieces that can be ground in this machine can only be round. It is possible to hand feed them. However, the machine can have an automatic tray feed which moves through a track and the work moves between the two wheels. The function of track feeding is to create similar parts of the same size. You can grind bolts with this machine if the diameter or outside diameter is set for grinding threads.

The centerless grinder has a work piece blade. This is a piece of steel that the parts rest on when going through and between the two wheels.

The following safety list is very important when using grinding machines:

- Check for cracked wheels.
- Never put a wheel that does not fit on the machine.

- Paper washers are used to give equal pressure on the wheel.
- When starting up a wheel remember to turn it on and off. Look at the wheel to make sure that it is running straight.
- Clamp work if it can not be held down on the magnet.
- On the magnet chuck, block the work so that if a large cut is made, the work will not move.
- Correct speeds and feeds on the machines.
- Remember to wear glasses.
- Correct wheel for the particular work at hand.

GRINDING WHEEL SELECTION

The first thing to know when selecting a grinding wheel is what kind of steel is being ground. This can be found on the blueprint. (*Blueprint reading can be found in chapter 4—Tool and Die Procedures*). Grinding wheels are made up of two basic elements; one is abrasive material and the other is a bonding compound used to hold it together. Grinding wheel composites are similar in structure to a handful of lathe cutting tools which are mixed together with a bonding material. The result is little points from the tool bits extruding out from the bond acting as cutting edges. Rotate this at a great speed and material will peel off as in a lathe and milling machine. Grinding will give off sparks when the wheel touches the work, which are particles of steel being removed from the work piece. A grinding wheel usually is able to sharpen itself. This means little pieces break off the wheel but it will build up or close up grains. This means your grinding wheel will not keep the size and start leaving burn marks on the part you are grinding.

Grinding wheels come soft, medium, or hard depending on what type of material is being ground. Wheel manufacturers can be helpful in developing new grinding methods and materials particularly when synthetic materials resist already existing machinery.

Grinding wheels come in many shapes. The straight grinding wheel comes in different sizes and is used on almost all grinding machines. Usually a three-fourth-inch thick straight wheel will be used on the surface grinder and a three-fourth- to two-inch will be used on cylindrical and centerless grinders. The cut-off wheel is a wheel that cuts steel off. Tube steel or cutting off a piece that you want a straight end on which is thin and brittle are examples of what can be obtained with this type of wheel. These wheels can come in sizes of 1/32 to 3/32 of an inch and are just used for cutting off material and only for that.

Recessed wheels are generally used in tool cutter grinders, and they have a recess in one side. The reason why they are recessed is so that they touch the least amount of steel with a small surface and there is less heat buildup. Dish and saucer wheels are used for the reamer or cutter that the straight or recess wheel cannot sharpen. The wheels appear and are shaped like a dish and a saucer.

The standard marking system for grinding wheels is as follows: 42 - A - 37 - M - 4 - B - 22. It is necessary here to discuss each letter or number in order to decipher its meaning. The number 42 represents the manufacturer's symbol indicting the exact kind of abrasive used in this particular wheel. The letter "A" will tell you the abrasive type that the manufacturer is putting into the wheel. It could be aluminum oxide or silicon depending on what letter the manufacturer is using. The number 37 indicates the grain and refers to the type of finish. Grain comes in four different types: coarse, where the number will be low, normally from 10 to 24, indicating that the grain or the stone content is far apart; medium-grain, where the numbers will be 30 to 60 with a closer stone structure; fine grain, where the number will be 70 to 180 with a very fine stone structure; and very fine grain, where the number will be 220 to 600. Coarse grain wheels are used for heavy cuts and for the work pieces that are big in diameter. Medium wheels are used for intermediate work which coarse and fine wheels either break down or build up. Fine wheels are used for fine smooth finish and small diameter work. The grade of the wheel is indicated by letters "A" through "Z" (soft, medium, and hard). Grade means the hardness of the wheel and how much bond material it has in it for resistance when grinding. Some toolmakers like grinding with hard wheels because the wheel will hold up and not break down. The number system 1–15 refers to the structure that is open and will take off more material. With a closer grain structure, grain will be used for fine finish work and closer tolerance work. Some toolmakers will sometimes keep the number on the wheel from about 5 to 7. The bond type which is indicated by letters is the material used in holding the wheels together. There are six standard bonds used and they have common, recognizable names. Rubber, for example, (R—rubber bonded wheels) is used extensively for wheels on centerless grinders. They produce a high finish and can be run at speeds up to 16,000 surface feet per minute. Shellac bonded wheels are designated by a letter "E." Wheels of this type are usually cutoff wheels which are very thin. Silicate bonded wheels are designated by the letter "S." These types of bonds are used mainly for large, slow rpm machines where a cooler cutting action is desired. If you are using a silicate wheel it should not run over 6500 surface feet per minute. Vitrified bond designated by the letter "V" is the most common bond used. Seventy-five percent of the wheels are made with this type of bond. The bond is not affected by oil, acid, or water. Rapid temperature changes in the wheel have little or no effect on the bond. One thing to remember is not to run the wheel over 6550 sfpm. Resinoid bonded wheels are designated by the letter "B." The resinoid wheels are strong and shock resistant and are made for cutting off steel. The machine that you would use is called a cutoff machine. The next and last item used in grinding wheels is oxychloride. This bond is generally used on grinding discs which are often used in auto body shops and when grinding steel with a hand grinder. The final number on the wheel is the manufacturer's code which is used to designate the wheel bond.

Diamond wheels are in a class by themselves. This wheel is used to grind carbide tool bits. A diamond wheel is used every day in the tool room by toolmakers. Their size varies from 1/32 to 1/14 of an inch.

The following safety list is very important when using the grinding wheel:

- Check for cracks.
- Ring the wheel by taping it with a piece of wood or plastic. Do not use steel. The ID (inside diameter) of the wheel you are ringing should be held with a piece of metal in order to get the ring effect. Using your finger does not work when listening for a crack.
- Wear glasses if you are near machines.
- Dressing of the wheels should be slow and never fast. The diamond dressers have been known to come apart and the diamond tips fly out of their holder.
- Many toolmakers have a habit of using any wheel, even if it is too large for the job. Try to avoid this.
- Keep the machine clean at all times.

THE JIG GRINDER

A jig grinder is used in precision grinding. This machine has speed ranges from 12,000 to 60,000 rpms which allow more accurate control of grinding and stock removal. It has a larger table size and is higher than other grinders. Its features are essential in the toolmaking business.

A simple, two-station die block is a typical example of the advantages of jig grinding operation. After a jig boring machine has bored the holes and contours, the piece is then heat-treated. After the heat treatment, the jig grinder is used to straighten out the distortion and to remove the decarbonization from the areas that will be used either as guides, punch reliefs, counterbored holes, blank holes, piercing holes, or angles. The machine has a head that moves up and down and turns clockwise or counterclockwise. This movement gives the finished part a smooth surface known as a micro finish.

Contour grinding is done on this machine with the help of a rotary table similar to the one mentioned in the milling section. Chop grinding provides a means for rapid stock removal in what may well be considered a vertically shaped movement with the wheel as a tool. Wipe grinding is sometimes used for finer finishes and is accomplished through a horizontal movement between the work and the wheel. This method is similar to surface grinding. Outfeed grinding is a method used in hole grinding whereby the material is removed through a continual outward movement of the grinding wheel. The machine can slot grind, blend in surfaces, grind radius, hide recesses, and sharpen corners. This machine is used to grind tolerances of aircraft parts that need a very close micro finish. Grinding wheels are either on a shaft from the manufacturer or are attached to an arbor. They are basically the same type of wheel used in surface grinding.

The rotary table is used for contour grinding and locating holes. The micro sine table is used in conjunction with the rotary table for perfect angle grinding. The table can be positioned from 0 to 90 degrees with the help of gage blocks. An angle plate can be used with the rotary table for 90-degree work. Parallel blocks are used for accurate set-ups from + or − .001. They are equipped with ground blocks, bolts, and height blocks. The gaging of holes can be measured with the help of the leaf taper gages. These gages size the hole you are bore grinding from .095 to 1.005 of an inch. They are ground within + or − .00025 of an inch.

The items that come with the jig grinder should never leave the machine or be used on any other machine in the tool room. This assures the accurate sizing for the jig grinder operation.

The following safety list is very important when using the jig grinder:

- Take care of the tool which is being used.
- Grinding wheel dust in the air is hazardous, therefore a mask should be used.
- Watch the high-speed rotation of the spindle.
- Optics should be clean at all times.
- Remember to tighten the grinding wheels.
- Watch the speed setting of the wheels.

BASIC METALLURGY

The term metallurgy is something that every toolmaker must know. Understanding the different properties in the metal is essential. Most metals that are used must go through structural changes. Brittleness, hardness, quantities of stress, and resistance to heat and corrosion must all be determined for accuracy. Fatigue in the material is the reaction which takes place in metal after stress has been applied. This is achieved by putting a drive shaft or a turbine shaft under a load. However, it will bend, break, twist, or seize if the direction is changed too fast.

The Brinell hardness test is one way to to check the hardness of the steel. This is used by the toolmaker to see if the metal was heat-treated correctly. A diamond is used to check the steel because it is the hardest stone and will penetrate hardened steel. The machine measures in pounds and they range in size anywhere from 1100 pounds to 10,000 pounds. The diamond will become larger every time you change the size of the steel because a bigger area needs to be checked.

PHASES OF HEAT TREATING

Acid brittleness is the brittleness induced in steel when it is pickled in a bath of a diluted solution of acid for the purpose of removing scale or the raised metallic rust. Acid lining is the inner bottom and lining of a melting furnace composed of materials that have an acid reaction.

Aircraft quality steel is steel that has been tested during manufacturing and has been approved as suitable for production of aircraft parts. Air hardening is a hardening process whereby the steel is heated to a hardening temperature and then cooled out in the air. Bend test is a test commonly made by bending a cold sample of a specified size through a specified circular angle. Bend testing provides information of the docility of the sample steel. Black annealing is a process of box annealing sheets prior to the tenning, or rust-protecting coating, whereby a black oxide color is imparted to the surface. A blow hole is produced during the metal process when gas is failing to escape, thus causing an air bubble. A blast furnace is a furnace supplied with air blast usually not enough to produce pig iron and accomplished by melting iron ore. Blue annealing is the process by which sheets are allowed to cool slowly after hot rolling. Blueing is a method by which sheets are coated. The blued surface is obtained by exposure to an atmosphere of dry steam or air at a temperature of about 1000° Fahrenheit.

Bright annealing is usually carried out in a controlled furnace atmosphere so that surface oxidation is reduced to a minimum and the surface remains relatively bright. A British thermal unit (BTU) is a unit of heat representing the amount of heat necessary to raise the temperature of one pound of water one degree Fahrenheit. Colorizing is a process of converting the surface of steel articles into a corrosion resistant alloyed layer of aluminum and iron. This is accomplished by the surface absorption of aluminum from a mixture of aluminum and aluminum oxide powder. Carburizing is the process by which carbon is added to steel-less alloys by absorption and by heating the metal at a temperature below its melting point when in contact with the material. A carburizing compound is a mixture containing carbon solids which will release carbon into steel in the presence of heat. Case hardening carburizing and nitriding is a subsequent hardening by heat treatment. All or part of the surface portions of a piece of iron release alloy.

Chemical analysis is a qualitative analysis which consists of separating a substance into its component elements and identifying them. Chipping is a method of removing surface defects such as small fissures. Chromium is a hard, corrosion-resistant metal widely used as an elemental alloy in steel and for plating steel products. Cold finishing changes the shape or reduces the cross section of steel while cold. Cold working permanently deforms metal below its temperature and this hardens the metal. Combined carbon is the permanent deformation of a metal below its recrystallization temperature, and this also hardens the metal. Controlled cooling occurs when cooling comes from elevated temperatures in a predetermined manner in order to avoid hardening, cracking, and internal damage, or to produce a desired microstructure.

PROCESSES OF HEAT TREATING

The following pages will introduce terms associated with the properties of materials.

Stress is the amount of force inside by which the material resists changing shape. Strain is a deformation or change in shape that is caused by applying a load. Strength is the property of a piece of steel that enables it to resist strain when stress is applied. Plasticity is the ability of a material to withstand extensive permanent deformation. It indicates a permanent change of shape in which it cannot return to its previous shape. Malleability is the property that makes it possible for material to be stamped, hammered, or rolled into shape or sheets. A malleable material is one that can withstand permanent deformation from compression. Lead is an example of a malleable material. Toughness is the quality that enables a piece of steel to withstand shocks and stresses and to become deformed without breaking. This means a piece of material can be bent one way and then the other without fracturing. Hardness of a piece of metal is defined as its ability to resist indentation, abrasion, or wear and cutting. The hardness of metals is usually associated with strength.

Fatigue is the action that takes place in a piece of metal after stress. When a sample is broken in a tensile test machine, a definite load is required to cause that fracture. Corrosion resistance is the ability of a piece of steel to withstand attack by the atmosphere, fluids, moisture, and

acids. Heat resistance is the property of steel that retains strength or hardness at high temperatures. A metal that retains its strength or hardness at elevated temperatures is called heat resistant.

Weldability and machinability are not strictly properties. They are important practical considerations in fabrication or repair of any metal part. Weldability is referred to as the ease with which a metal may be welded. Machinability is used to describe the ease with which metal may be turned, milled, or manipulated by any other machine operation.

Heat treatment is the operation or combination of operations that includes the heating and cooling of a metal in its solid state in order to develop or enhance a particular desirable mechanical property such as hardness, toughness, machinability, or uniformity of strength.

With regards to heat treatment, there is terminology that one should become familiar with. The following pages will introduce properties associated with heat treatment.

The chief purpose of annealing is to relieve internal strains and to make the metal soft enough for machining. The following list contains nine types of metals with annealing instructions:

- Cast iron requires heat slowing to between 1400° and 1800° Fahrenheit depending on the composition. Allow the cast to hold the temperature for about thirty minutes and allow the metal to cool in the furnace slowly.

- Stainless steel must be treated between 1850° to 2050°F for full annealing. Cool rapidly.
- Copper should be heated to 250°F and quenched in water. A temperature of 500°F will relieve most of the stress and strains.
- Zinc should be heated to 400°F, then cooled in open air.
- Aluminum should be heated to 750°F, then cooled in open air. This reduces hardness and strength but increases electrical conductivity.
- Nickel-copper alloys including monel should be heated between 1400° to 1450°F and cooled in water or oil.
- Nickel-molybdenum-iron and nickel should be heated between 2100° and 2150°F. This should be held at this temperature a suitable time, depending on thickness.
- With brass annealing, relieving stress may be accomplished at a temperature as low as 600°F. Brass should be slowly cooled to room temperature.

The following safety list is very important when heating metal:

- Never touch hot metal.
- When cooling parts in oil, follow procedures carefully.
- Be aware of temperature control.
- Note the color chart of the metal.
- Wear eyeglasses.
- Watch the temperature of the furnaces.

- Make sure the air flows through vents.
- Make sure the gas connections are tightened.

WHAT IS A ROBOTIC ARM?

The robotic arm was developed in industry for speed and safety. It moves on impulse and is used in conjunction with CNC computer controlled machines or assembly work and as a tool for the toolmaker. If used correctly, production is increased greatly in conjunction with a person.

Robots will play an important role in the industrial field but in no way threaten the job security of workers. There are more than thirty thousand robotics in industry. More than six thousand of them are in the United States where they are the working minority. For every robot in the United States there are more than forty thousand people. Each of these machines does a simple job like spot welding, spraying, unloading die castings machines, moving work from one machine to another, and changing tools. They do all this only when the toolmaker programs them to do so. The robotic arm frees the toolmaker to attend to other tasks.

Many toolmakers feel threatened by robotics because it is taking their jobs away. And yet the toolmaking field with the help of robotic arms will in fact get industry moving faster, and this will help toolmakers keep their jobs. The following are the common, Society of Automotive Engineers (SAE) steels to be used in making almost any part of a die. You should know that all metals have different properties, and you have to pick the right one for the job. Knowing the differences in the steels will help you determine what hardness and how much tension strength you will need for the die and the mating of parts. Steels are made for a specific job and that's why toolmakers have to know the difference between them and what is composed in that steel.

CHAPTER 2

THE HEALTH AND SAFETY ACT

GOVERNMENT, STATE, AND FEDERAL GUIDELINES

The Health and Safety Act can be found in volume 29 of the federal regulation code book which contains all of the federal regulations for all trades. The Health and Safety Act protects the worker and the company from any safety hazard.

The Codes of Federal Regulations have fifty volumes, of which Health and Safety is only one, with as much as 2,000 pages. The various volumes deal with everything from trade to office workers. A toolmaker should know the contents of eleven volumes as they cover office, labor, defense, shipping, employee benefits, banking, economic stabilization, accounts, internal revenue, public health, and pensions. Questions that deal with foreign shipments, pensions, and noise levels in the shop can be answered by reading these volumes.

WHAT IS OSHA?

OSHA (Occupational Safety and Health Administration) is a separate agency that protects the workers in all fields. The guidelines provide the worker and the employer with a safe working environment. OSHA standards should be followed at all workplaces. All companies should have an agency head that establishes and maintains on-the-job programs that meet OSHA requirements. The federal government passed OSHA laws in 1970. The federal government regulates OSHA programs and makes sure they work, and issue guidelines to set up safety and health programs. This government agency implements rules that assure workers that the work areas are safe and healthy at all times.

The company you work for should have the following:

- An appointed in-house official who works with OSHA

- An upgraded safety program

- Safe policies in the workplace

- Employees and employer to advise and assist in helping OSHA work

- OSHA standards and investigation of complaints from the employer and the employees

- Records on file for any accident or injury

- Adequate safety training programs for the management and employees

As a worker you should have a knowledge of the hazards in your workplace and cooperate with the inspectors from OSHA. Perform your job by following safety rules. Inform your supervisor if you notice any unsafe or unhealthy conditions. Report accidents and participate in OSHA programs. You can do a lot to make a safe work area. You have rights under OSHA as a worker. You have the right to be informed, to be represented, and to be protected.

However, some things are up to you. You should know your job and always be alert. You should inform supervisors of safety problems before they happen, and always use personal protection. You should also take first aid courses and take part in the company's safety programs.

YOUR OWN SAFETY PRACTICES

It is very important that you become safety conscious in the workplace. The first essential lesson to learn in the tool room is to work safely. You should always learn the safe way to do things and practice it every day for yourself and the workers around you in the tool room. These are standard rules that all toolmakers should follow in the workplace. As a toolmaker you should always wear safety glasses. This is for your protection from flying objects. Loose chains hanging off your neck, rings, watches, and loose clothing should be avoided. Sleeves on your long-sleeved shirts should always be rolled up when operating any machine. Apron strings, if you wear one, should be long enough so you can tie them in the back.

Files should be handled carefully in the tool room because they are one of the most common tools that can injure toolmakers. Do not play with metal chips. Shoes are a big factor in the tool room. Safety-tipped shoes should be invested in for your safety to prevent serious injury to your feet. Many shops and companies have their own safety programs. If you notice any safety hazards, you should bring it up to your supervisor. Remember, safety comes first. The biggest area of accidents are caused by rushing. Toolmakers and operators should not sit near their machine while it is running. It is also a bad habit leaving a job half done or leaving it with broken tools and never telling anyone about it, because this is how most of the accidents happen. Remember, a moment of carelessness may be a lifetime of grief for you and your co-workers.

CHAPTER 3

AREAS OF INSPECTION

WHAT IS QUALITY CONTROL?

Quality control is a source of inspection that follows the raw material to the final stage of a finished product.

Quality control begins when all raw material received is checked for damage and identified with color codes that the manufacturer uses to tell the type of steel you are using. However, most companies have their own code for the material coming in, and the toolroom foreman will most likely stamp the steel. Finished material and unfinished material are stored separately.

In-process inspection control occurs when the inspector is right there to check your work and stop you if its wrong. Final inspection must be approved before the shipping stage. Remember to work with the inspector.

KNOWLEDGE REQUIRED FOR INSPECTION

An inspector should know how a blueprint is made. He or she should understand the theory of projection drawings and have a knowledge of lines and their uses. Inspectors should understand the principle of orthographic projection which involves auxiliary views and sectional views. They should understand dimensioning on drawings, geometric and positional tolerancing, and metrics and metric drawings.

As a toolmaker or diemaker you will be working extensively with engineering drawings. All drawings have different shapes, dimensions, curves, straight lines, and figures that show the shape of the part in such a way that if you read and interpret the drawing you can make the object the way the designer intended it to be made. Most of the time you work directly with the designer and changes are made by both of you.

Blueprints are made from an idea or design that either the manufacturer or the designer draws up. The inspector must know the different lines and views to determine the method of inspection to be done. Many views are necessary for one part to be made. Some of the things the inspector must know are the part name, the material to be used, and the part number. Most inspectors have one job to do and that is to inspect the part you are working on only. These inspectors very rarely concern themselves with the assembly of that part. These are major problems with the industry today.

Because of the demand of quantity, quality is sometimes overlooked. This is one of the toolmaker's biggest problems in industry today. Toolmakers can visualize what they are making and what it will do when it is assembled. The inspector will only see the part they are working on at that one time. To solve this problem, both should work together and not contradict one another.

WHAT IS AN IN-PROCESS INSPECTION?

In-process inspection is when the inspector is on the line of production. This inspection is a control procedure with the intention to prevent the manufacture of unsatisfactory parts. It is desirable and to the toolmaker's advantage to obtain setup assurance and approval before having the operation continue. This inspection is usually the first step in overall inspection. An inspection form is made out at this point and follows the piece to the end of the operation. Inspection of parts produced in multiple spindles must be conducted consecutively by the same inspector. If the inspector and toolmaker are working together, the first piece passed will be tagged and put aside and then will follow the completed operation. This is done for the protection of the inspector, the toolmaker, and the operator. This preliminary inspection is also called the "first piece inspection," and it helps the toolmaker when that part comes around again. If there were no changes to the print, it should have no problem passing the inspection department. The in-process inspectors are supposed to follow the operations to the end and note any changes in the log. These logs are kept in the files daily and sometimes updated yearly.

FINAL INSPECTION

Each shop will have a final acceptance area for finished parts which is established in a controlled area with adequate lighting and temperature control. It is normally supervised by the manager of the inspection department. Final inspection must pass prior to any piece entering the stores or shipped to the customer. Final inspection occurs in a separate area from other departments. In some companies customers will have the final say on the parts. An approved sampling plan is usually on hand with special requirements that the final inspector will do.

The final inspection consists of a full check on all of the parts. The plan from the customers should have the lot size and the sample size. These will follow the finished part to the customer. Occasionally rejected parts will be sent along with the order and if changes in the prints were made they will also be sent along with the order. One of the reasons all work procedures must be logged is to track changes in the orders. Some companies may want the inspection files kept longer than one year. Companies that produce aircraft parts have been known to review logs ten years later. Material that is used must have the findings from the heat-treating company sent along with each order. A good toolmaker also makes a good inspector. This is why proficiency is necessary when inspecting a piece of material.

CHAPTER 4

TOOL AND DIE PROCEDURES

BLUEPRINT READING

Blueprint reading is necessary in the tool and die field. It is important to learn how to draw a blueprint from the beginning stages and to know how the print is set up with the different types of lines, symbols, and numbering systems. A toolmaker must know how a blueprint works and how to make changes in it. A blueprint is a type of map or chart on a piece of paper in which a designer indicates through standardized drawings what he or she would like made.

A blueprint is an engineering drawing, which is in essence a graphic form of communication from the designer to the toolmaker. The blueprint details what an object will and should look like. There are three steps that make up blueprints—the drawing, the pencil or ink tracing, and the print. On it are the lines, figures, and curves of the drawing. There is also a title block which will give the name of the print, the sizes to hold, angle tolerances, and who drew it. What I have taught many students to do, is draw the piece or part and make all the views on separate sheets of paper. At that point I have them tape all the views together in order to make a cube. They then look at it and open it up slowly while looking at all sides as they are laying the paper on a flat surface. By making a cube I would have them imagine the full scale of the part in the middle of the cube. When you draw up all different angles of a blueprint a toolmaker has to take those views and make that piece exactly as it was drawn.

An orthographic is when a designer makes a print with the basic three views and then puts in a third angle projection view. This third angle projection view gives you an idea of what the piece will look like. The third angle projection is usually drawn in because there is an angle or radius on the piece. Today blueprints are made on a blotting machine with the help of a CAD/CAM designer computer. Some of these CAD/CAMs have a mouse attachment, and if there is a change to the print the designer can change it on the screen, and even make another print.

Some of the things that are drawn in blueprints are drill sizes, reamed hole sizes, counterbore sizes, the depth of the hole, slots, grooves, and the finish that is required. Everything is in the working print to complete the job.

WHAT AREAS WILL BE COVERED IN BLUEPRINT READING?

Patience is required in blueprint reading. Many designers draw perfectly but without considering that someone else has to make this part. A good practice for a designer is to get involved in the machines that the toolmakers use and then draw with them in mind.

MATHEMATICS NEEDED

Basic math is necessary in the tool and die industry—such as addition, subtraction, multiplication, and division—along with trigonometry, geometric dimensioning, and tolerances.

The industry requires you to know weights, measures, symbols, decimals, fractions, millimeters, and how to convert fractions to metric sizes and much more. The mathematic formulas can be very difficult but necessary. With the help of good math instructors and updated material you can figure out any problem. Remember that math is very hard to understand for many people. However, if it is taught the right way there will be no time wasted in figuring out the problem, and the assembly line, presses, and mold machines will produce work smoothly.

WHAT KIND OF TOOLING IS MADE?

Tooling is made for many presses, and it ranges from a twenty-five-pound die to a four-thousand-pound die. Much depends on the size of the part or assembly. The die is built to make the part either finished or for some other operation.

The toolmaker has to come up with a design that will work for the part or parts like batteries, radiators, and window sections.

Toolmakers should know all areas of toolmaking. Someone who is simply building tools and no longer learning may have to change jobs to pick up a new skill in toolmaking. A good toolmaker is someone who can go into any shop, pick up a print, and build that part or die. If the work is the same thing every day for four years, you are not going to get the knowledge or experience in the toolmaking field that you need.

To continue in toolmaking you will find the following will give you an idea of some of the simple tools you will have to know and make at any time. Two items that you will start with will be a die set top and bottom. Either a two post or a three or four post will do. Each die is basically the same but some have more detail. There are punches, sleeves, stops, jam nuts, as well as the blanking punch, punch plate, pilot, stripper plate, automatic stop, finger stop, back gage, front spacer and die block.

Dies that use metal for making parts punch into it, and a design and complete part is made. A die for a mold machine is basically the same but you use plastic that is melted down in the machine and is forced into the die cavity, held under pressure and then released, and a mold comes out. It could be a flashlight casing, computer housing, printer housing, or anything with plastic. The plastic flows into the cavity and your two halves of the die will form the part you are making. With the tooling in the metal die each part in that die has to have a high-gloss finish which is sometimes polished for hours to get the right finish you need. This is the same with the mold dies. The polishing of the die is very important.

All the tooling that is made for these dies or molds will be made on lathes, milling machines, grinding machines, jig boring EDM machines, and the basic machines that you were taught when learning the toolmaking field.

Anything the die needs according to the print you will make from raw stock. Sleeves, bushing, and punches can be purchased locally. Because of the price factor and time involved in the assembly of the die almost all the pieces that are made from raw stock will have a standard to follow. This can be found in your *Machinery's Handbook,* which you should use every day. Or it can be found in the blueprint itself specifying the size they want.

You have to remember that tools cannot be made for the machines unless you have the background to figure it out.

DIFFERENT TYPES OF DIES AND MOLDS

There are many dies and molds assembly parts. There are blanking dies which blank out parts in three phases—accuracy, appearance, and flatness. Cutoff dies are for basic operations of cutting off stock. Compound dies produce a blank and may have pierced holes in them. Trimming dies trim off the excess metal to complete the finished part. Piercing dies pierce holes in a finished part that may need a certain number of holes put into that piece. Shaving dies remove a small amount of material from a blanked-out piece to give it the finished size. Broaching dies have a broach tool that cuts serrations into the piece using the push-down operation which opens the sections larger and larger. Horn dies take a piece of metal that is pre-formed and the die fastens it together like a round casing. Side CAM dies produce parts that are secondary which may call for a hole or slots in the side of the piece. Bending dies bend or deform pieces of metal to a desired shape. Forming dies are used to take a blank and form it into a curved contour or an angle with bends in it. Drawing dies are similar to forming dies, but they have a tighter closure tolerance when drawing. The amount of pressure has to be adjusted just right or the end drawing action will push out a blank slug. If you were drawing down a piece of steel to form a cup, the roundness would be there but the bottom would be punched out. Curling dies press the material down to form a shape like a cup shape but curl the edges similar to a sleeve with a flared-out corner shaped like a "T." Bulging dies form a piece of metal. The die has a formed top and bottom which creates a round radius. If you were to look at the piece it would look like a chef's hat. Extruding dies take a blank slug and force the metal down into a shape with angles and smaller diameters which are forced downward similar to a carbon dioxide container. Swaging dies create the opposite effect. The piece is preformed and shaped into angles and smaller diameters with the pressure being forced from the top of the die. With cold forming dies a blank or a slug of metal is put into the die and as pressure is put on the die, the metal is forced to move up or down to shape a part in the shape used in the beginning stages.

Progressive dies are when you feed a roll of sheet metal into the die and the piece is formed in different stages, starting from one end of the die to the other or halfway through the die. At this point a finished piece is completed and ready for inspection. Sub-press dies are used for small pieces similar to wristwatch parts. A piece of metal is fed in the die and the punching action of the top of the die hits one time and stamps out a finished piece. These dies can be operated at a great speed.

The assembly die assembles the parts. It assembles two or more at the same time—for example, a flat piece of metal pressing one or two pins into a prelocated hole in a plate or a flat piece of steel.

As a toolmaker you will come across some of these dies in the workplace. Almost all of the tool and die training programs have very knowledgeable teachers and the material to demonstrate it. The books on the market today have much information in them. They will help you acquire a knowledgeable background in toolmaking. As a toolmaker, you will need to have an answer for any problem that arises on a machine operation, and it must be figured out in a short amount of time.

RECOMMENDED READING

Libraries will have the basic toolmaking books. There are also school or apprentice programs that give you a wide range of learning material. Math books, trigonometry formulas, charts, and blueprint design booklets will provide some information. Most important is the toolmaker's bible. It is a book that toolmakers and machinists consistently follow. This book is called *Machinery's Handbook*. This book will be used daily in your career as a toolmaker. This book has everything that is needed to know in the field, such as sizes, threads, tapers, formulas, allowances tolerances, feeds, hardness testing, metals, different fittings, surface texture, gear ratios, bolts, bearings, strengths of material, mathematical tables, and much more. If you are interested at this point, talk to someone, preferably in the toolmaking field. Talk to someone who is retired from the trade. Pick up material such as machine design books or manufacturing design books. You can become a good toolmaker with the right education. Some toolmakers start small, at a small shop, then progress to a larger company.

CHAPTER 5

TOOL AND DIE MAKERS

The progressive toolmaker has to be very proficient operating the large presses. They create many pieces per minute. The difference between the progressive toolmaker and other toolmakers is a closer observance of the presses. Many operations are performed in order to make a completed piece. Every operation has to be checked and approved by the inspection department. There can be over fifty different operations to complete one whole piece. Progressive toolmakers work long and frustrating hours to complete one good sample and must adjust the die and the press often.

PROGRESSIVE TOOL AND DIE MAKERS

A progressive die maker is a person who can take a blueprint of a part and design a die. The toolmaker's knowledge can create anything for that die. With a background of math, blueprint reading, and layout, the toolmaker can decide what material and machine is required. There are fourteen areas of a die that the toolmaker has to take into consideration before he or she begins designing the die. These steps are very important when making a die. Following methods that are taught to you by experienced people will help you achieve the best working die set possible. Read the blueprint and see if the scrap piece that the die will produce when in operation can be made. For example, if a piece is to be punched out with holes on it, it is necessary to draw the part and draw the scrap piece in order to see what you are going to make before you start. Getting lost in a blueprint is very easy. Once you know how the scrap piece will look, you can begin on the die block which will punch out the piece and the holes on the blanking punch and the piercing punches. Remember, all this time you are doing this on paper. The punch plate is a plate that secures the piercing punches together. You have to know what kind of pilots to use. A pilot holds the piece in line. You are at the point of needing to know what size, material, punches, and die blocks to use but you must first have a gage in front and back of the strip of material in order to have it in the center of the die block. Stops have to be considered to stop the strip of material to the size location of the piece you want. Stripper plates are now included to prevent the piece from coming up with the punch or punches. The stripper is fastened into the die block. The next thing is to find the best size die set and mount it in a press. Simple dies can become very hard or very simple. All this time you have been drawing this piece and all the parts for the die set. If you made your dimensions, notes, and a paper sample, you should be able to build it out of metal now. Pencil and paper should be your beginning stages of designing any die. Make a mock model of the piece and see what it looks like before you waste metal, time, and money for the company your work for. All I have done is shown you the difficulty of making a die. Starting with a pencil and a piece of paper is the right way to start. Really know the piece you are making, get involved with it and make it. There is much involved in designing a die which requires a lot of training.

The following is a very important safety list when working as a progressive tool and die maker:

- Be careful lifting heavy dies; some weigh up to fifteen hundred pounds.
- Make sure the two parts of the die are apart. Do not work between them.
- Always have help in removing any heavy part of the die.
- Always keep your hands out of the area of the die when it is in operation.
- If you are setting a large press make sure the safety devices are working.
- Many of the presses today have coolant or hydraulic systems, and when they fail the machine may not stop and may keep running on. Noises should be checked if they do not sound normal.

MOLD MAKERS

Mold makers have to know about the temperature of the material going into the mold. This material is little pieces of plastic that melt under extreme temperatures. Some molding methods are: compression molding, transfer molding, injection molding, cold molding, continuous extrusion, and molded laminated. The molding materials that you will come into contact with are: phenolic plastics, amino plastics, cellulose plastics, ethyl cellulose, acrylic plastics, vinyl plastics, styrene plastics, nylon, shellac and cold molded plastics. Plastic product design consists of moldability, parting lines, knockout pins, pickups, shrinkage, dimensional tolerances, warpage (which there is a lot of when the material is too hot or too cold), draft, wall thickness, fillets, sharp corners, ribs, bosses, holes, molded threads, inserts, lettering on mold products, and surface finish. A mold construction can consist of mold plates, sprues, runners, gates, cavities, venting, mold pins, ejector pins, return pins, guide pins, guide pin bushings, steam lines, and assembly of all of the items mentioned above. The material consists of low carbon steel, medium carbon steel, high carbon steel, alloy steel, beryllium copper, melanite, and bronze material. Before making the mold there are some elements that should be checked: the location of the knockout pins, the location of the parting lines, the location of the gates, the location of the pickups, whether or not the inserts should be molded in or pressed in, what wedges are required, the location of the radii, whether or not the cavity should be milled, whether or not the cavity and plunger should be made in one piece, and what kind of finish is required. These are some of the things that must be decided when you are making a mold. Mold machines have tons of pressure. When in operation the force of these machines is enormous. So remember as a mold maker keep your ears and eyes open.

The following safety list is very important if you work as a mold maker:

- Check the air pressure.
- Check the heaters in the machine.
- Check the flow of material in the melted state.
- Check the overflow or bridges to make sure they are in the correct location; otherwise, they will build up pressure and push the mold apart.
- Make sure the gates in the mold are not clogged.
- Make sure the safety devices are set properly.
- Always remember the molds are very heavy and you should always have help when moving, replacing, and taking them apart.

EYELET TOOLMAKERS

An eyelet toolmaker is not someone who makes sewing needles. That is a different type of toolmaker. Eyelet machines have clustered heads which vertically point down to a flat table surface. This table is where the dies are fitted and the stock runs over them. The heads move downward from left to right increasing the size of material to form a part. The heads that toolmakers must make are made in different sizes in order to form the part to the exact dimension. Examples of work that an eyelet machine makes are battery cases, lipstick tubes, and anything that would be drawn down to different sizes at a great speed. The machines go so fast that you cannot count the number of pieces they are putting out. This is just like the large ton presses that the progressive toolmaker sets up. The difference between the eyelet toolmaker and other toolmakers is that they make tools that look like pins. These pins are made up of a very hard steel, sometimes made by using carbide. Carbide tools last long but break very easily so caution has to be taken when making and installing the tools. The eyelet toolmakers must be as knowledgeable as the other toolmakers and they still have to go through the tool and die apprentice program. They have to know how to run all types of lathes, millers, and grinders and be able to perform heat treating, jig boring, benchwork, EDM honing, lapping, brazing, and sometimes welding. Eyelet toolmakers spend about three thousand hours on the floor troubleshooting. The biggest part of the training for eyelet toolmakers is to learn the machine operation and to keep their ears open for noises that will indicate the machine is having a problem.

The machines have a coolant that is used to help the piece draw down and have a finish that is passable to the inspector. The benchwork that you will come across will be small tools to larger ones. Filing and polishing is the most important thing that the eyelet toolmaker does. The micro finish that must occur on the tools should be smoother than glass.

The following is a very important safety list when working as an eyelet toolmaker:

- Hands should be under the punches when the machine is on.
- Avoid moving the sheet metal by hand when the machine is in operation.
- Check the safety devices and micro switches.
- Check the coolant flow and the last time the coolant was changed.
- Dispose of the coolant in a Department of Environmental Hazards (DEH) approved container because the coolant gets contaminated from use.

CHAPTER 6

BECOMING A SUCCESSFUL TOOL AND DIE MAKER

SCHOOLING

It is important that high school or trade school be completed. If you have attended high school look for a night education program that offers programs in the tool and die field. Look for a company with in-house training or a company that will work with a night education program. Contact your state job service and look into the programs that are available and whether they are accredited through your state. The next important factor is the basic knowledge of blueprint reading. It is important to know how to draft a simple sketch or blueprint of a small part or object and to know how to read a blueprint. You must know what sectional drawings are, with their different views and the symbols that are used in and on blueprints.

You must know the basic math that is needed including fractions and decimals. Geometric construction, graphs, and charts are important to know for the interview. It is also good to have a basic knowledge of the lathe, milling machine, drill presses, and CNC.

TRAINING

Training is available in every state. In fact all states and Puerto Rico offer programs. Most states have standard programs where companies train apprentices. In doing so the company is then open to grants from the state and federal government. Many companies allocate many opportunities for the toolmakers for advancement.

Try to find a company that offers good benefits and has a good working environment in which to learn. Ideally, companies with apprentices would like them to stay in the company. Apprentices can transfer to another company if they feel they are not getting proper training.

Look for the apprentice councils in the state in which you live. Look in chapter 8 under State Apprenticeship Agencies and you may find there is a company looking for apprentices.

LENGTH OF TRAINING

If and when you decide to become a toolmaker apprentice, the time involved in the program will feel like years of training. Eight thousand hours is the total amount required. Roughly five hundred hours are involved in theory and six thousand hours in the workplace. This does not mean that you will stay for the eight thousand hours with no credit of any kind. When you are evaluated by the company and the school that will be teaching the theory, you will find out that many of the hours will be credited to you for your knowledge and your schooling. If you took courses in any technical colleges, or took classes in a trade school that was state controlled, and attended ninth through twelfth grade in a state-run trade school, you may find many credits in your favor.The state will comply with giving the apprentice credit hours for on-the-job training, trade schools, and schooling related to the trade. Holding a high grade average while in school can also be credited toward theory.

Toolmaking for some companies is a way to teach apprentices the business. In becoming a toolmaker, you must find a reputable company. Make sure to have a signed agreement explaining what you will learn, what your pay will be to start, what your benefits will be, and what process will allow you to obtain your tool and die papers.

Apprentices start at the bottom of the pay scale. You must prove yourself and show that you are eager to learn. Remember that having a signed agreement is very important. Make sure you read it carefully. Call the State Apprenticeship Council in your area to make sure the company is accredited and approved by that state.

WHAT IS IN-HOUSE TRAINING?

To start in-house training the company must have an approved in-house apprenticeship training program. The company will also offer theory courses in house. Many have state instructors or college instructors come to the company and teach the theory in a classroom. Some of the ground rules that an in-house apprentice training program should have are that each apprentice will inevitably move at his or her own pace. Accomplishments in advancement of the state license level (eight thousand hours) will be recognized through the company's own journeymen's certificate. This means that if the company has its own theory courses in-house taught and approved by the state, the company can give any amount of hours to the apprentice that it would like. All the state requires is that you complete the eight thousand hours in the program. Many companies will subtract from the total of the eight thousand hours needed to complete the program the hours that the employee has worked on some of the machines. Theory taught in-house can be done in one year if you maintain a high average and attend three classes a week that are three hours a day. Otherwise theory, at a less accelerated rate, takes four years to complete.

Theory projects include safety, making a threading tool, turning tools, and facing tools as well as turning, facing, knurling, and threading a project. For the accelerated class, the first three months have projects that involve center punch, spring center pin, edgefinder, and flycutter blank. Most of the first three months are involved with pedestal grinders and lathes. The next three months are projects like size blocks, parallel bars, parallel clamps, tap wrenches, completing the flycutter blank, small V-block set, and a sine bar. The machines involved in those next three months include surface grinders and bridgeport millers, and techniques include sawing, heat treating, finish grinding, and lapping of the pieces. The last three months involve projects with depth gages, screw thread gages, large V-blocks, C-clamps, tap wrenches, center tester, center punches, and boring bores.

There are specific requirements that must be met to be an apprentice. Apprentices must be eighteen years old and pass tests of measuring skills and ability levels. They must be tested with the Thurston's Mental Alertness Test—a battery of tests given to measure motor functions, mobility, and mental aptitude—as well as on reading and arithmetic skills and with the California Personality Inventory. Candidates also have an evaluation of relevant experiences, attitudes, motivation, and educational background. Potential toolmakers who pass these tests are assigned to what tool section they want to be in, and a toolmaker is assigned to them as an assistant instructor to help them through the projects and machine operations. After every one thousand hours of work they are taken in front of a board of toolmaker supervisors and questioned in the area they are learning. In-house training moves at a fast pace. The company will need these toolmakers as soon as they are released from the program.

THE QUALIFICATIONS FOR APPRENTICESHIP

If you are picked by a company to work as an apprentice, there are some qualifications and standards that you have to understand. A typical company which has the apprentice program will have standards already set up. The following is a list from one of the big manufacturing companies:

- The apprentice shall mean a person selected by the company council to learn the trade and who has signed an apprenticeship agreement.
- Apprenticeship agreement shall mean a written agreement between the company and the apprentice.
- The apprenticeship council shall mean that group which has been designated to administer the apprenticeship training program in the company.

The following shall constitute the membership:

- The manager or a delegate chosen by him to be his representative.
- The supervisor of personnel of industrial relations.
- The training manager or coordinator.
- The supervisors selected from the tool rooms and/or other persons sufficiently qualified in apprenticeship training to take an active part and who have been approved by the council.

Qualifications also include:

- Graduation from an accredited high school trade school or technical college or other qualifications such as experience or relative training which, in the judgment of the company or the company's council, shall be deemed equivalent.
- Applicants should have average or higher grades in mathematics.
- The applicant shall be physically capable of performing the work of the trade. A certificate from the company or an outside doctor will be required. The reason for this is to make sure that the apprentice has no oil reaction, metal reactions, or physical handicap that may interfere with becoming a toolmaker apprentice.
- The applicant must be morally and mentally sound.
- To aid in determining his or her ability to work successfully with others, the applicant will be required to undergo suitable tests which have been approved by the apprenticeship steering committee.

Look at Appendix F for a sample you may be given when being interviewed for the apprenticeship program. This may help you understand what some companies are looking for in apprentices.

CHAPTER 7

FIELDS OF EMPLOYMENT

MAJOR INDUSTRIES IN THE UNITED STATES

It is the large industries that will most likely put ads in the newspapers. They include United Technology, Aircraft, UTC Pratt, Whitney Aircraft, General Motors Corporation, which has an apprenticeship program, American Standard, Stanley, and many more. The best way to find a company or a major industry is to work through the guidance office of your school or call the nearest state apprenticeship council for help. Most of the state apprenticeship offices have a listing of major industries that offer the apprenticeship program. If you are unemployed, there are agencies that can be contacted, and they will find you a position or an interview with companies offering programs. Your best bet is to start at the apprenticeship agency in your state.

TEACHING TOOL AND DIE

Teaching tool and die is not an easy job. You can teach it two ways. One is teaching it in a trade school, and this will be only twelfth-grade students. The other is to teach in-house training of tool and die apprentices in a company. This includes teaching theory, math, and possibly on-the-job training. Many companies that have the apprenticeship programs now use teachers from outside the company for their apprentices. Some of the colleges have programs for blueprint, math, and machine technology that will be a part of the company training programs in that area. The company signs an agreement with the school and its teachers to put together a program for the tool and die apprentices. Some of the states have been working very closely with companies trying to help the students graduate with updated knowledge of the new technology.

Remember that if you were a tool and die instructor, you would have to have had roughly eight to twelve years of working in the trade. The field is very large and there are many different toolmaking skills necessary in order to be a good teacher. A good teacher is someone who has a correct answer for a student or apprentice at all times. Apprenticeship programs are equal opportunity programs.

WHAT DO I NEED TO TEACH TOOL AND DIE?

Schools that you will apply to as a teacher will require you to have had courses in supervising so that you can supervise your students. You must take courses in forming your own teaching outlines and courses in discipline so that you will handle different situations when they come up. Most of the states are beginning to test teachers and instructors in their trades to see if they are keeping up with the new technology.

You should remember that you will be teaching either high school or beyond high school level. Some of the students in the high school will challenge you in your knowledge, and you have to know how to handle that.

WHAT ABOUT YOUR OWN BUSINESS?

Did you ever think about starting your own business? There are some things that you should know before you decide to open one. The first is to find a good location. Check to see if there are any small companies that want to sell. Make sure you have enough money to start one, and a knowledge of contracts and contacts in the right companies. Decide whether or not you want a partner to join with you.

The next thing to do is to find a backer. This is someone or a group of people who are interested in supporting your business. You will need money to start. If you are fortunate to have the funds to start, then you are on your way to starting your own business. There are, as well, grants in the federal government that help small businesses get started. The grant can be given to you free or you might have to pay it back, depending on what kind of grant you get. Remember that anyone can start a business. It can take up to five years to see if there is any profit.

The following list consists of some things you may need to think about before starting your own business:

- Do I have enough knowledge to run my company?
- Do I have enough contacts out in industry?
- Do I have enough capital to start?
- Do I need partners?
- Can I run the company on my own or do I have to hire employees?
- Do I have a good location for the business?
- Do I want to have all my time dedicated to the business?

How Do I Start My Own Business?

First travel to small and large companies to see for yourself how business are faring. Go to the large companies to see if any of them are contracting out work to small businesses. There are also salespeople who can help you get work for a 5 to 15 percent commission.

Some of the questions you may ask yourself about running an independent business are the following:

- Can I get the work out on time?
- How long will I have to wait for the money?
- Can I cover my expenses between payments from jobs contracted out which are not paid as frequently as an employee paycheck?
- Will I be able to stay ahead in making parts and shipping them out on time?
- Will I go into debt by borrowing money to cover expenses?
- Will the company make money in the long run? Are more loans necessary to keep the business going?

Estimating Jobs

In order to run a business successfully it is necessary to know how to estimate jobs.

There are several things to consider. Direct cost means the cost directed to the part, subassembly, and product you are making. It includes the cost for the tooling, and whether they have to be made for the job, and the cost of metal. You can round this up with two considerations. One is to figure out how the cost is directly related to the manufacturing of the product. The other is how practical it is to relate the cost to a particular item. Indirect costs cover those items necessary to operate the manufacturing section but are not traceable directly to one specific part. You have to remember that a machine breaking down, utility bills, and materials and tooling are all indirect costs. Indirect costs are broken down further in your company with the burden of overhead. A general rule of cost estimating is that it is concerned with the company's overhead but not with the general, overhead expenses. Company overhead may include all costs associated directly with the operation of your company, but not directly to a product. Factory overhead can be cutting oil for the tooling lights, heating for the company, and any labor costs that might be involved.

Standard versus actual cost is when manufacturing costs are classified as either standard or actual. Standard cost is an ideal cost. The variances between actual costs and standard costs provide you with a tool to evaluate the effectiveness of your working force, which means that you found a way to cut costs in the time it takes to manufacture the part and have less tool breakage.

Cost responsibility falls under all aspects of the company's expense. You should always have a set price when you figure the job and make sure you are not going to lose any capital. This price should make you profit in the long run.

Cost reduction includes cost control, which reduces the cost of manufacturing the part while maintaining the quality of the work. Value analysis optimizes value added to a part by reducing manufacturing costs. Value engineering goes beyond cost control and represents an attempt to combine two functions in both long- and short-range planning. Cost estimating is an attempt to predict the costs that must be incurred to manufacture the part.

Cost estimating is important to any company. A carefully prepared estimate is important when deciding whether to begin manufacturing a part. Good cost estimates may be used to:

- Establish the bid price of a part for quotation or contract.
- Verify quotations submitted by you from the vendor.
- Provide information for make-or-buy decisions.
- Make sure you have the most economical method and process or material for manufacturing the part.

There are two types of estimates. First there is a preliminary estimate for new parts before designs and plans are completed. This means that you may have to rotate the methods of operation that you have planned for this part. This is not the final cost. If blueprints are not available it is very hard to determine the actual cost of making the part.

A final estimate and detailed estimate should be based on all plans and manufacturing costs of making the part. This includes tooling metal and machine time.

Some do's and don'ts of cost estimating:

- Do use a printed form wherever and whenever possible, to assure completeness.
- Do have files on all estimating of parts to refer to in the future.
- Don't make an estimate on a part that needs a lot of outside inspection reports without including them.
- Do keep searching for new material and easier ways to make the part.

- Do keep a sample and a copy of the print on file to refer to at a later date.
- Don't guess at something you do not understand when estimating; call the vendor and ask about it. One thing to remember is to make all costs based on all operations and do not leave any operation out of the estimate. You can lose a lot of money if you make an estimate and then find out that you did not figure in the shipping cost.

You can see that estimating is a field of its own. You may want to take a course in a college on estimating.

UNIONS IN INDUSTRY

Unions in industry are a great benefit to have when you are working in a large or small industry. However, while working as an apprentice you may not qualify for the same benefits as regular workers.

If difficulties arose, the apprentice would be laid off first. Unions represent the employees who have their toolmaker's papers and employees who will be able to vote for the union.

When looking for work in a large or small company, check to see if the union is represented there. You should meet with them and see if they have an agreement that will protect you from being laid off as an apprentice. Although under union representation, all employees have equal rights, apprentices will often have less power than workers with their toolmaking papers even though they are paying dues.

Apprentices are sometimes caught in the middle of union and company problems.

Apprentices are the lowest-paid, and they are not considered skilled workers. Your seniority and paying dues does not count when voting. An apprentice is not a toolmaker until he or she has papers signed by the state.

In one company, apprentices had to belong to the union but still were expected go to work even if a strike was on simply because the contract of the company stated that all apprentices worked under the administration's power and were not allowed to be involved with strikes involving the union. This obviously created a tense situation for the apprentice.

There are unions and companies that work together very closely to help all employees, including the apprentice, to get better health insurance, more holidays, and good working conditions.

The following is what a typical union will do if collective bargaining is needed for the employees' benefit. If their contract is up in six months, the company and union must meet to do collective bargaining.

The union representative will collect all employees' complaints and suggestions and present them at the meeting. It may look like this:

- A three-year contract.
- A request to keep the union in the company.
- A list of dues and new employee fees.

- Grievances on all machinery that is outdated and needs repair.
- A company wide seniority list on job bidding and on layoffs.
- A 7 percent general wage increase, plus a cost of living increase effective January 1 of each year.
- Elimination of the bonus system in the assembly area and a 25-cent-per-hour increase in place of the bonus.
- A twelve-day holiday plan with pay.
- Forty-five hours of vacation after one year of service, ninety hours after two years of service, 120 hours after five years of service, 240 hours after fifteen years of service.
- A pension plan to be paid for from one part by the employees and the other part by the company after thirty-five hours of work and after five years of service.
- Full company insurance on all employees after ninety days as long as the number of employees does not decrease lower then the prescribed number outlined in the insurance contract.
- Sick pay to employees of $100 a week, if employed three years.
- A $2,000 life insurance policy for every employee, male or female.
- A 25-cent-per-hour increase for night shift workers.
- Automatic wage increases for new employees after ninety days and up to one year, then reverting back to the wage policy.

- The use of all bulletin boards and a 10-minute wash-up time before and after each shift.
- Two breaks, one from 10 to 10:15 A.M., the other from 2 to 2:15 P.M.
- Time-and-a-half for all Saturdays and for all hours over forty.
- Apprentices to have seniority after two years of employment and have their choice of shifts.

CAREER RESOURCES

TWO-YEAR COLLEGE PROGRAMS
IN TOOL AND DIE

Alabama

Muscle Shoals State Technical College
    Muscle Shoals, AL 35662
    205-381-2813

Arizona

Mesa Community College
    Mesa, AZ 85202
    602-461-7000

California

Chabot College
    Hayward, CA 94545
    415-786-6714

Long Beach City College
    Long Beach, CA 90808
    213-420-4135

Rancho Santiago College
    Santa Ana, CA 92706
    714-667-3016

San Bernardino Valley College
    San Bernardino, CA 92403
    714-888-6511

Georgia

Gainesville College
    Gainesville, GA 30503
    404-535-6241

Illinois

Black Hawk College, Quad Cities
    Moline, IL 61265
    309-796-1311

College of Lake County
  Grayslake, IL 60030
  708-223-8800

Kankakee Community College
  Kankakee, IL 60901
  815-933-0242

Oakton Community College
  Des Plaines, IL 60016
  708-635-1700

Prairie State College
  Chicago Heights, IL 60411
  708-709-3516

Triton College
  River Grove, IL 60171
  708-456-0300

Indiana

ITT Technical Institute, Evansville
  Evansville, IN 47715
  812-479-1441

Vincennes University
  Vincennes, IN 47591
  812-885-4313

Iowa

Hawkeye Institute of Technology
    Waterloo, IA 50704
    319-296-2320

Iowa Western Community College, Council Bluffs
    Council Bluffs, IA 51501
    712-325-3277

Marshalltown Community College
    Marshalltown, IA 50158
    515-752-7106

Northeast Iowa Community College
    Calmar, IA 52132
    319-562-3263

Kansas

Allen County Community College
    Iola, KS 66749
    316-365-5116

Pittsburg State University
    Pittsburg, KS 66762
    316-231-7000

Michigan

Macomb Community College
Warren, MI 48093
313-445-7999

Wayne County Community College
Detroit, MI 48226
313-496-2500

Minnesota

Alexandria Technical College
Alexandria, MN 56308
612-762-0221

St. Paul Technical College
St. Paul, MN 55102
612-221-1370

Wilmar Technical College
Wilmar, MN 56201
612-235-5114

Mississippi

Itawamba Community College
Fulton, MS 38863
601-862-3101

Northeast Mississippi Community College
   Booneville, MS 38829
   601-728-7751

Northeast Mississippi Community College
   Senatobia, MS 38668
   601-562-3200

Missouri

Maple Woods Community College
   Kansas City, MO 64156
   816-436-6500

Nebraska

Southeast Community College, Milford
   Milford, NE 68405
   402-761-2131

New Jersey

Passaic County Community College
   Paterson, NJ 07509
   201-684-6868

New York

Rochester Institute of Technology
   Rochester, NY 14623
   716-475-6631

North Carolina

Alamance Community College
Haw River, NC 27258
919-578-2002

Asheville-Buncombe Tech. Community College
Asheville, NC 28801
704-254-1921

Caldwell Community College & Technical Institute
Hudson, NC 28638
704-728-4323

Central Carolina Community College
Sanford, NC 27330
919-775-5401

Fayetteville Technical Community College
Fayetteville, NC 28303
919-323-1276

Wake Technical Community College
Raleigh, NC 27603
919-772-7500

Wilson Technical Community College
Wilson, NC 27893
919-291-1195

North Dakota

North Dakota State College of Science
  Wahpeton, ND 58075
  701-671-2201

Ohio

Ownes Technical College
  Toledo, OH 43699
  419-666-3282

Sinclair Community College
  Dayton, OH 45402
  513-226-2963

Pennsylvania

Johnson Technical Institute
  Scranton, PA 18505
  717-342-6404

Pennsylvania College of Technology
  Williamsport, PA 17701
  717-327-4761

South Carolina

Florence-Darlington Technical College
  Florence, SC 29501
  803-662-8151

Tri-County Technical College
Pendleton, SC 29670
803-646-8361

Tennessee

Chattanooga State Technical Community College
Chattanooga, TN 37406
615-697-4401

Wisconsin

Milwaukee Area Technical College
Milwaukee, WI 53233
414-278-6370

Moraine Park Technical College
Fond du Lac, WI 54935
414-929-2122

Western Wisconsin Technical College
La Crosse, WI 54601
608-785-9476

VOCATIONAL SCHOOLS

Arkansas

Black River Vo-Tech
Pocahontas, AR 72455
501-892-4565

Foothills Vo-Tech School
 Searcy, AR 72143
 501-268-6191

Pines Vo-Tech School
 Pine Bluff, AR 71603
 501-535-6054

Pulaski Voc-Tech School
 North Little Rock, AR 72118
 501-771-1000

California

National Technical School-Los Angeles
 Los Angeles, CA 90037
 213-234-9061

Connecticut

Porter & Chester Institute-Enfield
 Enfield, CT 06082
 203-741-2561

Porter & Chester Institute-Stratford
 Stratford, CT 06497
 203-375-4463

Indiana

Interstate Technical Institute
  Fort Wayne, IN 46803
  219-749-8583

Kansas

Manhattan Area Voc-Tech School
  Manhattan, KS 66502
  913-539-7431

Maine

Central Maine Technical College
  Auburn, ME 04210
  207-784-2385

Minnesota

Anoka Technical Institute
  Anoka, MN 55303
  612-427-1880

Dakota County Technical College
  Rosemount, MN 55068
  613-423-8301

Dunwoody Institute
  Minneapolis, MN 55403
  612-374-5800

Hennepin Technical College-Brooklyn Park Campus
  Brooklyn Park, MN 55455
  612-425-3800

Hennepin Technical College-Plymouth
  Plymouth, MN 55441
  612-559-3535

Rochester Technical College
  Rochester, MN 55904
  800-247-1296

St. Paul Technical College
  St. Paul, MN 55102
  612-221-1300

Staples Technical College
  Staples, MN 56479
  218-894-1168

Wilmar Technical College
  Wilmar, MN 56201
  612-235-5114

Missouri

St. Louis Technical
  St. Ann, MO 63074
  314-427-3600

New Jersey

Camden County Voc & Tech School
  Sicklerville, NJ 08081
  609-767-7000

Ohio

Akron Machining Institute
  Barberton, OH 44203
  216-745-1111

ITT Technical Institute-Youngstown
  Youngstown, OH 44501
  216-747-5555

Pennsylvania

Johnsonville Technical Institute
  Scranton, PA 18508
  717-342-6404

Puerto Rico

Escuela Voc de Area-Bayamon
Bayamon, PR 00619
809-785-2388

Escuela Voc de Area-Ponce
Ponce, PR 00731
809-842-7091

Rhode Island

New England Institute of Technology-Warwick
Warwick, RI 02886
401-467-7744

Tennessee

State Area Voc-Tech School-Crossville
Crossville, TN 38555
615-484-7502

Wisconsin

Acme Institute of Technology-Hales Corners
Greenfield, WI 53220
414-281-2111

Acme Institute of Technology-Manitowoc
Manitowoc, WI 54220
414-682-6144

REGIONAL APPRENTICESHIP PROGRAMS

Location	States Served
Regional Director Region One 11th Floor One Congress Street Boston, MA 02114	Connecticut, New Hampshire, Maine, Rhode Island, Massachusetts, Vermont
Director Region Two Federal Building New York, NY 10014	New Jersey, New York, Puerto Rico, Virgin Islands
Director Region Three Gateway Building Philadelphia, PA 19104	Delaware, Virginia, Maryland, West Virginia, Pennsylvania

Director
  Region Four
  1371 Peachtree Street
  Atlanta, GA 30367

Alabama,
Mississippi, Florida,
North Carolina,
South Carolina,
Georgia, Kentucky,
Tennessee

Director
  Region Five
  230 South Dearborn
  Chicago, IL 60604

Illinois, Indiana,
Michigan, Ohio,
Minnesota, Wisconsin

Director
  Region Six
  Federal Building
  525 Griffin Street
  Dallas, TX 75202

Arkansas,
Oklahoma, Texas,
Louisiana, New
Mexico

Director
  Region Seven
  Federal Office Building
  Kansas City, MO 64106

Iowa, Missouri,
Kansas, Nebraska

Director
  Region Eight
  United States Custom House
  19th Street
  Denver, CO 80202

Colorado, South
Dakota, Utah,
Montana, North
Dakota, Wyoming

General questions regarding programs can be addressed to:

United States Department of Labor
Employment and Training Administration
Bureau of Apprenticeship and Training
Francis Perkins Building
200 Constitution Avenue N.W.
Washington, DC 20210
(202) 535-0540

STATE APPRENTICESHIP AGENCIES

The following are state apprenticeship agencies. They can help you find companies or main manufacturers in your area that would have an approved apprenticeship program. If you are a senior in high school, you can be advised by your guidance office. Some schools have a manufacturing representative who visits the school for students who are interested in the manufacturing area.

Arizona

Apprenticeship Services
Arizona Department of Economic Security
438 West Adams
Phoenix, AZ 85003

California

Division of Apprenticeship Standards
  3950 Oyster Point
  Center Wing
  San Francisco, CA 94080

Connecticut

Office of Job Training and Skill Development
  Connecticut Labor Department
  200 Folly Brook Road
  Wethersfield, CT 06109

District of Columbia

District of Columbia Apprenticeship Council
  500 C Street N.W.
  Washington, DC 20001

Delaware

Apprenticeship and Training Section
  Division of Employment and Training
  Delaware Department of Labor
  State Office Building
  820 North French Street
  Wilmington, DE 19801

Florida

Bureau of Apprenticeship
Division of Labor, Employment and Training
Department of Labor and Employment Security
1320 Executive Center Drive Atkins Building
Tallahassee, FL 32301

Hawaii

Apprenticeship Division
Department of Labor and Industrial Relations
830 Punch Bowl Street
Honolulu, HI 96813

Kansas

Kansas State Apprenticeship Council
Department of Human Resources
401 SW Topeka Boulevard
Topeka, KS 66603-3182

Kentucky

Apprenticeship and Training
620 South Third Street
Louisville, KY 40202

Louisiana

Louisiana Department of Labor
  5360 Florida Building
  Baton Rouge, LA 70806

Maine

Bureau of Labor Standards
  State House Station #45
  Augusta, ME 04333

Mariana Islands

Apprenticeship Training Council
  Guam Community College
  P.O. Box 23069 G.M.F.
  Guam, Mariana Islands 96921

Maryland

Apprenticeship and Training Council
  1100 North Utah Street
  Room 310
  Baltimore, MD 21201

Massachusetts

Department of Labor and Industries
   Division of Apprentice Training
   Leverett Sallonstall Building
   100 Cambridge Street
   Boston, MA 02202

Minnesota

Division of Apprenticeship
   Department of Labor and Industry
   Space Center Building, 4th Floor
   443 Lafeyette Road
   St. Paul, MN 55101

New Mexico

Apprenticeship Bureau
   Labor and Industrial Division
   New Mexico Department of Labor
   501 Mountain Road N.E.
   Albuquerque, NM 12240

Nevada

Nevada State Apprenticeship Council
   505 East King Street, Room 601
   P.O. Box 4452
   Carson City, NV 89710

New Hampshire

New Hampshire Apprenticeship Council
   19 Pillsbury Street
   Concord, NH 03301

New York

New York State Department of Labor
   State Office Campus
   Building 12, Room 140
   Albany, NY 12240

North Carolina

North Carolina Department of Labor
   4 West Edenton Street
   Raleigh, NC 27601

Ohio

Ohio State Apprenticeship Council
   2323 West 5th Avenue, Room 2140
   Columbus, OH 43216

Oregon

Apprenticeship and Training Division
Room 404, State Office Building
1440 Southwest 5th Avenue
Portland, OR 97201

Pennsylvania

Apprenticeship and Training
1303 Labor and Industry Building
7th and Forster Street, Room 1303
Harrisburg, PA 17120

Puerto Rico

Incentive to the Private Sector Program
P.O. Box 4452
505 Munioz Rivera Avenue
San Juan, Puerto Rico 00986

Rhode Island

Rhode Island State Apprenticeship Shore Council
200 Elmwood Avenue
Providence, RI 02907

Vermont

Apprenticeship and Training
  Department of Labor and Industry
  120 State Street
  Montpelier, VT 05602

Virgin Islands

Division of Apprenticeship and Training
  P.O. Box 890 Christiansted
  Department of Labor
  St. Croix, Virgin Islands 00802

Virginia

Division of Labor and Industry
  P.O. Box 12064
  Richmond, VA 23241

Washington

Department of Labor and Industries, ESAC Division
  General Administration Building
  MS HC-730
  Olympia, WA 98504-0631

Wisconsin

Department of Industry, Labor and Human Relations
Employment and Training Policy Division
P.O. Box 7972
Madison, WI 53707

BASIC METALS

A. Kinds of Steel

The free running steels by SAE number:

1112
1113
1115
1120
1212
1213

Carburizing steels are listed in two categories:

Carbon Steels	Alloy Steels
1020	3115
1015	3120
1010	5120
1025	2317
	4615
	4620
	8617

Heat treating steels, there are two categories, rolled and annealed, for machinability.

Rolled	Annealed
1030	1045
1035	4345
1040	4340
1045	1050
1050	4130
3130	3130
3135	4140
2330	3135
3140	4150
2345	3140
	2330
	3145
	5140
	3150
	6150
	9260
	5150
	2340
	2345
	1095
	52100

B. The SAE System of Steel Classification

The following should give you a basic idea how to read the codes of steel.

The main subdivisions of the system:

1. carbon steels
2. nickel steels
3. nickel-chromium steels
4. molybdenum steels
5. chromium steels
6. chromium-vanadium steels
7. tungsten steels
8. silicon-manganese steels

Practice should be provided on the identification of the various steels showing how the number is analyzed, as in the example of 2345 SAE steels

Nickel		Points of Carbon
2	3	45
	% of nickel	

In the case of the tungsten high-speed steels, 71360 being typical, it is important to remember that tungsten alloys in larger percentages than nickel and that the amount of tungsten is indicated by the second and third figure.

Tungsten		Carbon Content
7	13	60
	% of tungsten	

OSHA REGIONAL OFFICES

United States Department of Labor regional offices for the Occupational Safety and Health Administration

Region I (CT, ME, MA, NH, RI, VT)

16-18 North Street
One Dock Square, 4th Floor
Boston, MA 02109
617/223-6710

Region II (NY, NJ, PR, VI)

Room 3445, One Astor Plaza
1515 Broadway
New York, NY 10036
212/944-3426

Region III (DE, DC, MD, PA, VA, WV)

Gateway Building, Suite 2100
3535 Market Street
Philadelphia, PA 19104
215/596-1201

Region IV (AL, FL, GA, KY, MS, NC, SC, TN)

1375 Peachtree Street, N.E.
Suite 587
Atlanta, GA 30367
404/881-3573

Region V (IL, IN, MN, OH, WI)

230 South Dearborn Street
32nd Floor, Room 3263
Chicago, IL 60604
312/353-2220

Region VI (AR, LA, NM, OK, TX)

555 Griffin Square
Dallas, TX 75202
214/767-4731

Region VII (IA, KS, MO, NE)

911 Walnut Street
Kansas City, MO 64106
816/374-5861

Region VIII (CO, MT, ND, SD, UT, WY)

Federal Building
1961 Stout Street
Denver, CO 80294

Region IX (CA, AZ, NV, HI)

Box 36017
450 Golden Gate Avenue
San Francisco, CA 94102

Region X (AK, ID, OR, WA)

Federal Building
909 First Avenue
Seattle, WA 98174
206/442-5930

BLUEPRINT TERMINOLOGY

Ang A geometric form that two lines meet at one point.

Assy A mechanism consisting of two or more parts placed in proper location.

Aux An orthographic view not contained in any of the six regular planes of projection, but constructed from one or more views.

Aux View A view pulled off to the side separate from the main view. This is done to highlight an important element to the part you are making.

Br A nonferrous metal that is an alloy of copper and zinc color which ranges from yellow to red.

C to C Center to center, a line that runs between and intersects the center of two features.

Centerline Two lines crossing each other.

Ch   The hardness of the outer layer of a part. To harden only the outer surface of a part.

Cham   An angle cut across the edge of a part to give a circular finish look, and remove all sharp edges of the surface.

Dimension Line   A line from one end to the other with the size in the middle.

Dimensioning on Drawings   There are four basic types: in line dimensioning, base line, dimensioning systems, and drawing scales.

Extension Line   A line extending off the end of the exact size of the part or at both ends of the dimension line.

Fin   Any final surface preparation on a part that protects its surface finish, or micro finish.

Hidden Line   Something that is hidden.

LH   Left-hand view.

Mach   To perform a machining operation of grinding, milling, or lathe work drilling.

MS   Machine steel free cutting.

NC   National course thread.

NEF   National extra fine thread.

N.P.T.   American National pipe thread.

Object Line   The darkest line on a blueprint.

Screwthread Lines   Long and short lines going downward.

Section Line   A line splitting the part through two areas of the view to show you the inside of the operation that has to be done.

Sectional Views   Consist of about ten different types: full section, half sections, broken out sections, rotated sections, removed sections, assembly sections, offset sections, angular sections, sectional symbols, detailed sections.

Special Dimensions   Examples include: diameters, radii, angles, chamfers, limits, allowances, holes and threads, thread notations, and tapers.

Tir   A total indicator with the highest to the lowest reading.

MATH EQUATIONS

The required positional tolerance formula

F = maximum diameter of fastener

H = minimum diameter of clearance hole

T = positional tolerance diameter

$T = H - F$

Tapped Holes Location

$$T = \frac{T_1 + T_2}{2}$$

Formulas for Tension and Compression

A = area of cross section of material in square inches

a = area of web of I-beam in square inches

E = modulus of elasticity in pounds per square inch

I = moment of inertia of section about an axis passing through the center of gravity in inches to the fourth power

M = maximum bending moment in inch-pounds

P = total tensile or compressive stress (load) in pounds

y = distance from center of gravity to most remote fiber in inches

Ss = stress in tension or compression in pounds per square inch

Z = section modulus for bending in inches to the third power

v = total shearing load at a given section in pounds

e = elongation or shortening in inches

l = length in inches

Formula for Tension and Compression

$$S = \frac{P}{A}$$

$$e = \frac{P_1}{AE}$$

Formula for Shear

$$Sa_{\text{ave}} = \frac{V}{A}$$

Formula for Bending

$$S = \frac{My}{I} = \frac{M}{Z}$$

Formula for Checking Screw Threads

$$\text{SIN } a = \frac{A}{B - A}$$

A = difference in diameters of the large and small wires used

B = total difference between the measurements over the large and small wires

a = one-half the included thread angle

Example: The diameter of the large wires used for testing the angle of a thread is 0.116 inch and of the small wires 0.076 inch. The measurement over the two sets of wires shows a total difference of 0.122 inch instead of the correct difference, 0.120 inch, for a standard angle of 60 degrees when using the size of wires mentioned.

The amount of error is determined as follows.

$$\text{SIN } a = \frac{0.040}{0.122 - 0.040} = \frac{0.040}{0.082} = 0.4878$$

APPRENTICE INFORMATION AND FORMS

SELECTION CRITERIA FOR APPRENTICE TOOLMAKING

You may run across a selection board and the board may have the following point system outline set up for them to fill in and you to answer, just like a job interview.

Part–1

1. Education background
   High school or equivalent (5) _____

   Beyond high school formal
   study 1/yr (5) _____

Subject courses related to apprenticeship training. (3 pts/course completed within 4 years)   _____

2. Previous experience
    Work experience pertaining to job-related apprenticeship training
    (2 pts/year experience)   _____

3. Seniority (2 pts/year up to maximum of 10 pts)   _____

4. Personal interview
    Can-do, will-do, how do factors
    (0–28 pts)   _____

    Job history (0–24 pts)   _____

5. Comprehension testing
    (0–28 pts)   _____

    (% of score × 28 pts)   TOTAL   _____
6. Thurston's Mental Ability   _____
7. California Personality Inventory   _____

Part–2

Job training
Present job title or company   _____

8. Personal Interview of   _____

·

On each of the following criteria mark one and only one description.

Poor = 1 point   Above average = 3 points
Average = 2 points   Excellent = 4 points

1. Health and energy   _____
2. Intelligence   _____
3. Motivation   _____
4. Maturity   _____
5. Outgoing   _____
6. Enthusiastic   _____
7. Ability to get along and working
   with others   _____

      Subtotal   _____

Job history

8. Attendance (lost days since hired) _____
9. Adherence to plant rules and concepts
   (disciplinary action if any)   _____
10. Attitude on present job (if employed)
    Other companies _____
11. Previous work performance   _____
12. Responsibility to job, self, others _____
13. Adherence to safe work practices   _____

      Subtotal   _____

      Total   _____

Interviewer _____   Date _____

APPRENTICE TRAINING PROGRAM

Method: The program will start with current company employees. Any who qualify will enter the program. They will be scheduled according to rating and be started at various starting points.

They will be expected to accomplish a certain workload concurrent with their training. This will be organized by the supervisors and trainer constantly.

One of the measurements will have to do with a value system. Attendance, work habits, productivity will all assist in the acceleration process.

Note: It is the intention of the company to recognize . . .

a. Those toolmakers who exceed normal skill levels and help them expand on their talents.

b. The need to develop a tradition, once strong in the old guild concept, that cultivates the same pride in craftsmanship and sustains the company.

Accordingly, the company:

1. will pay instructors to teach.
2. will pay apprentices to learn rapidly.
3. will reward accomplishments.
4. will provide the necessary tooling and equipment to train with.

Facilities Organization: The (company) Apprentice Training Program has been organized to . . .

1. Accelerate the development of qualified toolmakers for the projected journeyman needs high-tech training during its hi-tech growth period.

2. Reward those who meet the requirements of each phase by increasing their pay level as each stage is accomplished to the new measurements established . . . particularly if achieved at a greater speed.

3. Accomplish within the company facility . . .

   a. with an assigned toolroom trainer.

   b. with a classroom instructor from a state approved facility

   c. with paid time for shop and classroom hours (not overtime except where specially authorized by management)

d. with space and equipment assigned

e. with a broad spectrum of skill and experience emphasized to increase the versatility of those involved

f. with the involvement of supervisors as part of a review board to advise and assist the trainers and training manager

    — performance standards
    — use of equipment
    — discipline
    — curriculum
    — progress of program

g. the review board will meet monthly under the direction of the training manager with training staff present.

4. The program must recognize that . . .

a. Every apprentice starts at a deserved level based upon . . .

    — review data at start of program
    — test results
      . . . measurement of real value of hours to date
      . . . measurement of skills and abilities levels (psych. testing)
      . . . rating of instructors
    — state and companies bench marks.

b. Each apprentice is different and will move at his/her own pace and must be so developed.

c. It will be the serious obligation of accelerating this progress as quickly as possible . . . imposing greater loads continually.

d. Accomplishments in advance of the state license level (8000) hours are recognized toward a journeyman certificate only. The state license still requires 8000 hours.

The state requirement of 560 hours can be recognized "in equivalents" for classroom work. The instructor will recognize all of the training accomplishments.

e. For accomplishment of the total curriculum (of 8000 hours) in less than 4000 hours . . . a *one time* bonus of $_____ will be awarded.

   — also in less than 5000 hours $_____
   — also in less than 6000 hours $_____

There will be eight (8) stages. . . .

STAGE I — 1000 HOURS

   PRACTICAL _____

   CLASS ROOM _____

Upon completion (in whatever length of time) pay level will increase $_____ or _____ %.

STAGE II _____ HOURS

    PRACTICAL _____

    CLASS ROOM _____

Upon completion — pay level will increase
$_____ or _____ %.

STAGE III _____ HOURS

    PRACTICAL _____

    CLASS ROOM _____

Upon completion — pay level will increase
$_____ or _____ %.

STAGE IV _____ HOURS

    PRACTICAL _____

    CLASS ROOM _____

Upon completion — pay level will increase
$_____ or _____ %.

STAGE V _____ HOURS

    PRACTICAL _____

    CLASS ROOM _____

Upon completion — pay level will increase
$_____ or _____ %.

STAGE VI _____ HOURS

PRACTICAL _____

CLASS ROOM _____

Upon completion — pay level will increase

$_____ or _____ %.

STAGE VII _____ HOURS

PRACTICAL _____

CLASS ROOM _____

Upon completion — pay level will increase

$_____ or _____ %.

STAGE VIII _____ HOURS

PRACTICAL _____

CLASS ROOM _____

Upon completion — pay level will increase

$_____ or _____ %.

Finally: Upon completion of the (company's) Toolmaker Program . . .

a. The participant will be called a Journeyman Toolmaker.

b. He/she must keep a separate log for the state license and complete 8000 hours.

c. If he/she leaves the company before the receipt of his/her certificate (except for bad health or retirement) he/she will receive a certificate of hours.

d. He/she will be known as a Journeyman Toolmaker 'B' on completion of Step VIII.

Upon completion of another test, he/she can advance to Toolmaker 'A' and an attendant wage increase.

Note: It will be the intention to develop a "Master Toolmakers" category for those who attend a "Refresher Course" and meet a set of advanced standards.

This group will receive attendant wage recognition and a certificate.

The Master Toolmakers will be designated Toolmaker 'AA.'

GLOSSARY

A

Acid A large class of substances, the aqueous solution of which are capable of turning litmus indicators red or reacting with and dissolving certain metals to form salts.

Arbor An axis or shaft supporting a rotating part on a lathe, a bar for supporting cutting tools.

B

Bond A substance or an agent that causes two or more objects or parts to cohere.

Bore To make a hole in or through as with a drill, boring bar, or endmill.

Brinell hardness The relation hardness of metal and alloys determined by forcing a steel ball into a test piece

141

under standard conditions and measuring the surface area of the resulting indentation (BHN).

Burr   A rough edge or area remaining on metal or other material after it has been cast, cut, drilled, milled, or ground.

C

Center   A point equal distance or the average distance from all points on the sides or outer boundaries of something (middle).

Center drill   A bit having a point of a drill and a counter sink with the standard angle of 59 degrees (for the drill to follow in the location of the center drill).

Chamfer   A flat surface made by cutting off the edge or corner of something, such as a bar of stock turned in a lathe or a block of metal milled in a milling operation.

Chuck   A clamp that holds a tool or the material being worked on in a machine such as a drill or a lathe.

Compound angle   Two or more planes at different angles.

Contour   The outline of a figure, body or mass. A line that represents such as outline.

Coolant   An agent that produces cooling, a fluid that draws off heat.

Countersink   To enlarge the top part of a hole so a screw or bolt head will be flush on or below the surface of whatever part you are countersinking.

Cylindrical   Having the shape of a cylinder, especially of a circular cylinder.

D

Diamond   An extremely hard, highly refractive, colorless, or white crystalline of carbon.

Die   Any of various devices used for cutting out, forming, or stamping material. A metal block containing small conical holes through which plastic, metal, or other stock is extruded or drawn.

Dovetail   A fan-shaped tenon that forms a tight interlocking joint when fitted into a corresponding mortise to combine or interlock.

Drill   An implement with cutting edges or a pointed end for boring holes in hard materials machine operated.

E

Extruding   To shape (metal or plastic) by forcing through a die thrushing out.

F

Flowchart   A schematic representation of a sequence of operation.

G

Grinder   A person who sharpens cutting edges and performs the operation of grinding something.

H

Hardness   The relative resistance of a mineral to scratching, and resistance of a metal to denting, scratching, or bending.

Helical   Having a shape of a helix.

Helix   A three-dimensional curve that lies on a cylinder or cone and cuts the elements at a constant angle.

I

Industry   The commercial production and sale of goods and services, a specific branch of manufacture and trade.

Inspection   Official examination or review.

Internal   Relating to or located within the limits or surface of something inside.

Intersection The point or locus of points common to two or more geometric figures.

K

Knurling One of a series of small ridges, as along the edge of an object such as a thumbscrew.

L

Lathe A machine on which a piece of metal or other material is spun and shaped by a fixed cutting or abrading tool.

M

Machine finish A finish on metal surfaces.

Metalurgy The science or procedures of extracting metals from ores, of purifying metals, and of creating useful objects from metals.

Mill Machines for shaping, cutting, polishing or dressing metal surfaces.

Mill finish A smooth surface on various metals made by a machine.

Q

Quench To cool (metal by thrushing) in water or other liquid.

R

Radius A line segment that joins the center of a circle with any point on its circumference.

Ream To form, shape, taper or enlarge (a hole) with a reamer, to remove material.

Resinoid Pertaining to or containing resin, a resinoid synthetic.

S

Shank A piece of metal or other material used to reinforce or shape this part.

Speeds The rate of a measure of the rate of motion.

Stagger To place regularly on alternating sides of a midline set in a zig-zag row.

Stainless Any of various steels alloyed with sufficient chromium to resist corrosion, oxidation, or rusting.

Stress An applied force or system of forces that tends to strain or deform a body.

Subdivisions The act or process of subdividing.

T

Tooling   The process of providing a factory with machinery in preparation for production.

T-slot   A milled slot fixed in a machine surface shaped like a "T" for square-head bolts.

Turning   To cause material to move around a center or focal point, to rotate or revolve.

U

Undercut   A cut to create an overhang by cutting material away.

V

Vise   A clamping device of metal usually consisting of two jaws or more which may be closed or opened.

Vitrified   Through heat fusion.

VGM CAREER BOOKS

VGM Career Horizons
a division of *NTC Publishing Group*
4255 West Touhy Avenue
Lincolnwood, Illinois 60646-1975